T0185573

SpringerBriefs in Plant Science

SpringerBriefs present concise summaries of cutting-edge research and practical applications across a wide spectrum of fields. Featuring compact volumes of 50 to 125 pages, the series covers a range of content from professional to academic. Typical topics might include:

- A timely report of state-of-the art analytical techniques
- A bridge between new research results, as published in journal articles, and a contextual literature review
- A snapshot of a hot or emerging topic
- An in-depth case study or clinical example
- A presentation of core concepts that students must understand in order to make independent contributions

SpringerBriefs in Plant Sciences showcase emerging theory, original research, review material and practical application in plant genetics and genomics, agronomy, forestry, plant breeding and biotechnology, botany, and related fields, from a global author community. Briefs are characterized by fast, global electronic dissemination, standard publishing contracts, standardized manuscript preparation and formatting guidelines, and expedited production schedules.

More information about this series at http://www.springer.com/series/10080

Girdhar K. Pandey • Swati Mahiwal

Role of Potassium in Plants

 Springer

Girdhar K. Pandey
Department of Plant Molecular Biology
University of Delhi South Campus
New Delhi, Delhi, India

Swati Mahiwal
Department of Plant Molecular Biology
University of Delhi, South Campus
New Delhi, Delhi, India

ISSN 2192-1229   ISSN 2192-1210   (electronic)
SpringerBriefs in Plant Science
ISBN 978-3-030-45952-9   ISBN 978-3-030-45953-6   (eBook)
https://doi.org/10.1007/978-3-030-45953-6

© The Author(s), under exclusive license to Springer Nature Switzerland AG 2020
This work is subject to copyright. All rights are reserved by the Publisher, whether the whole or part of the material is concerned, specifically the rights of translation, reprinting, reuse of illustrations, recitation, broadcasting, reproduction on microfilms or in any other physical way, and transmission or information storage and retrieval, electronic adaptation, computer software, or by similar or dissimilar methodology now known or hereafter developed.
The use of general descriptive names, registered names, trademarks, service marks, etc. in this publication does not imply, even in the absence of a specific statement, that such names are exempt from the relevant protective laws and regulations and therefore free for general use.
The publisher, the authors, and the editors are safe to assume that the advice and information in this book are believed to be true and accurate at the date of publication. Neither the publisher nor the authors or the editors give a warranty, expressed or implied, with respect to the material contained herein or for any errors or omissions that may have been made. The publisher remains neutral with regard to jurisdictional claims in published maps and institutional affiliations.

This Springer imprint is published by the registered company Springer Nature Switzerland AG
The registered company address is: Gewerbestrasse 11, 6330 Cham, Switzerland

Preface

Rapid changes in the environment at the global level affect crop productivity. A plant must contain an appropriate amount of nutrients in order to sustain under extreme environmental conditions. The famous trio of nitrogen, phosphorus, and potassium (shortened as NPK) act as a safety gear for plants to withstand harsh environmental conditions. Due to their utmost importance in the growth and development of plants, it is imperative to understand the functions mediated by these nutrients at the cellular level. In this book, role of potassium (K^+) has been briefly discussed, from its importance in agriculture to plant development and stress tolerance.

K^+ plays a key role in all forms of life, starting from a single-cell bacterium to highly complex organisms such as plants and animals. However, the K^+ content in each organism is quite variable, probably due to the unique cellular requirements and diversity in function. It is most abundantly present in plants among all the life forms on Earth. Talking about its importance, one can say that some of the roles played by K^+ are basic in origin as they are required in the basic functioning of a cell.

Universal roles of K^+ have been described briefly in Chap. 1. Apart from its role as a nutrient, it is a major intracellular ion required for osmotic balance in a cell. Since it is important for osmotic balance in a cell, cells must maintain K^+ homeostasis in order to maintain cell structure and turgidity throughout their life. In Chap. 2, a brief glance of homeostasis mechanism in bacteria, yeast, animals, and plants have been discussed. K^+ homeostasis in a cell is maintained by well-defined transporters and channels. Chap. 3 places a spotlight on K^+ transport systems in plants and discusses briefly the K^+ channels and transporter families in plants along with their regulation. K^+ transport system shows a high level of conservation among all kingdoms, which have been briefly discussed in Chap. 4. They are known to have common characteristic features and similarity in structure, especially in the pore region. Evolutionary conservation among plant channels and transporters has been determined, showing their overall conservation in the plant kingdom as well as how they have diverged so far from their last common ancestors.

For plants, the extreme importance of K^+ can be understood by its crucial role in plant development. Though well-known for its role in opening and closing of stomata, it is required at almost every stage of development as well as growth. Chap. 5 provides a broader view of the role played by K^+ in plant growth and development. In their entire life cycle, plants often encounter precipitous abiotic stresses such as drought, waterlogging, low temperature, and high salinity. Chap. 6 reflects on all these types of abiotic stresses providing a clear picture of interplay of K^+ in these stresses. Another challenge faced by plants is the constant change in nutrient availability that depends on the availability of nutrients in the soil. As K^+ is very limited in the soil, there are likely chances of plants facing K^+ deficiency. Chap. 7 provides brief insight into K^+ deficiency, explaining common symptoms of K^+ deficiency that are quite visible in plants. K^+ deficiency acts as stress signal in plants to switch on various mechanisms, enabling cellular machinery to replenish overall K^+ availability in plants. The overall mechanism revealing how K^+ deficiency is actually being sensed in the plants is yet to be deciphered. Speculations have been made that the response to K^+ deficiency is initiated at the roots. Chap. 8 highlights the various possible mechanisms that have been suggested to sense K^+ deficiency in plants.

Besides all the aforementioned functions of K^+ in cells, several other roles of K^+ in plant such as biotic stress responses have now come into the limelight. These newly emerged roles have been described in Chap. 9. But it needs to be emphasized that these roles need to be established well with strong evidence. Nevertheless, more experimental proofs might soon connect plant K^+ homeostasis with biotic stress resistance. Also, the concept of of K^+ as a second messenger is a challenging issue because of its limited properties that do not go well with the characteristic features required for a second messenger. This needs to be further investigated by experiments with solid proof. Chap. 10 deals with several key questions and future perspectives in this area. In summary, it can be concluded that K^+ is a versatile and indispensable nutrient required by plants for their proper growth and development.

New Delhi, India   Girdhar K. Pandey
                                                                                Swati Mahiwal

Acknowledgments

We are thankful to the Department of Biotechnology (DBT), Science and Engineering Research Board (SERB), Council for Scientific and Industrial Research (CSIR), and Delhi University (R&D grant), India, for supporting the research work in GKP lab. We acknowledge Department of Biotechnology (DBT), India, for providing DBT-Junior Research Fellowship to SM. We also express our thanks to Mr. N. Pavithran and Dr. Sibaji K. Sanyal (Department of Plant Molecular Biology, University of Delhi South Campus, New Delhi, India) for critical reading of this manuscript.

Contents

Chapter 1
Role of Potassium: An Overview

Introduction

One-third of global gross domestic product (GDP) was a contribution from agricultural sector in the year 2014, hence declaring this as a crucial parameter for commercial growth. In a country like India whose GDP is majorly based on the agricultural sector, maintenance of crop health as well as its nutritional value is a challenging task for farmer. If one has to ask an Indian farmer the secret of their agricultural productivity, nitrogen-phosphorus-potassium (NPK) fertilizer wouldn't be a new word. This signifies how important NPK is, for a farmer, for a country and for the crops to meet increasing demand of food. These three essential macronutrients play major role in plant growth and development.

In all organisms, from prokaryotes to eukaryotes, K is a major constituent and key nutrient required by all forms of life. It is usually required in large amounts by plants and animals for their proper growth and development. Animals obtain adequate amount of K from their food, directly from the plants and indirectly from animal products. Since plants are the major dietary source of K for animals, in order to meet all the requirement of K by animals, plants must contain more amount of K compared to animals. This may be one of the reasons that it is abundantly present in the plants.

K comprises about 2.6% of earth's crust, hence ranked as the seventh abundant element on earth (Wedepohl 1995; Schroeder 2019). The composition of K in the soil solution is very important since soil solution plays a critical role in determining the K uptake by the roots (Leigh and Jones 1984). The four major K pools that have been recognized in the soil are (i) soil solution K, (ii) exchangeable K (EK), (iii) slowly exchangeable or fixed K (SEK) and (iv) structural or unavailable K. EK and soil solution K are in dynamic equilibrium; hence, EK is also considered to be available to plants (Römheld and Kirkby 2010). Large reservoir of total K is present in the soil, but K availability for the plants is very limited because EK and SEK are the

© The Author(s), under exclusive license to Springer Nature Switzerland AG 2020
G. K. Pandey, S. Mahiwal, *Role of Potassium in Plants*, SpringerBriefs in Plant
Science, https://doi.org/10.1007/978-3-030-45953-6_1

major contributors of K uptake in plants. These two pools (EK and SEK) only make up to 1–2% and 1–10% of total K, which is usually located at the topsoil. Out of all four forms, EK form is the only actively available form for the plant, and its measurement can reflect soil K status readily available for plants (Leigh and Jones 1984; Römheld and Kirkby 2010).

In plant tissues, K is the most abundant cation, which exists as a free ion, i.e. (K^+). K^+ has an essential role in cell metabolism, growth and development as young developing tissues and reproductive organs contain higher concentrations of K^+ (Mengel and Kirkby 2001; Amtmann et al. 2008). At physiological level, long-distance transport of K^+ is essential for neutralizing inorganic and organic anions, acting as dominant cation in xylem and phloem. In plants, almost 6% of dry matter is K, an indicative of high abundance in plant (Leigh and Jones 1984). Inside plant cells, the concentration of K^+ in cytosol is up to ~200 mM, and it is required in the transport of organic anions and metabolites, maintenance of transmembrane voltage gradients for cytoplasmic pH homeostasis, protein synthesis, photosynthesis, osmoregulation and cell extension (Jeschke et al. 1997; Rüdiger and Ralf-Rainer 1997; Mengel and Kirkby 2001).

Importance of Potassium

Potassium as a Nutrient

K's role as a nutrient is well established even before the twentieth century in animals and plants. In animals at physiological level, it is involved in acid-base regulation, nerve and muscle function, kidney health, bone health, blood pressure and heart function. In plants, it regulates several physiological processes such as building and strengthening of the plant, improving movement of photosynthates in the plant, enhancing resistance to pests and diseases as well as regulating water status and increasing drought tolerance in the plants (Zorb et al. 2014). Inside the cells, it is responsible for ion homeostasis, protein synthesis, osmoregulation, enzyme activation, regulating membrane potential and charge balance in both animals and plant cells (Fig. 1.1) (Egilla et al. 2001; Umar and Moinuddin 2007; Marschner 2012).

Potassium: Major Intracellular Ion

Being one of the most abundant cations in plant cells, a relatively constant concentration of 80–100 mM is maintained in the cytoplasm along with accumulation of large pool of K^+ inside vacuole (Dreyer and Uozumi 2011). However, K^+ concentration in the cytosol of various organisms differs significantly. In *Escherichia coli* K^+ concentration is maintained at 250 mM in the cytoplasm during normal growth con-

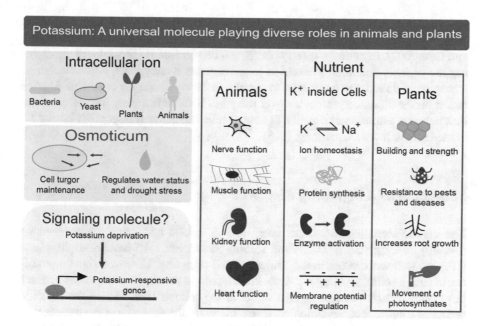

Fig. 1.1 Schematic representation of diverse roles of K in animals and plants, where it is responsible for the regulation of array of responses at cellular and physiological levels. K is a major intracellular ion acting as a nutrient, osmoticum and signaling molecule in animals and plants

ditions, whereas *Saccharomyces cerevisiae* cytoplasmic concentration of K^+ ranges from 200 to 300 mM depending on the strain and its growth conditions (Navarrete et al. 2010; Ashraf et al. 2016). Protozoa, such as *Tritrichomonas foetus*, has up to 119 mM of constantly maintained intracellular concentration of K^+ (Maroulis et al. 2003). In fungus, *Dendryphiella salina* protoplasmic K^+ concentration is reported to be 30–60 mM. In invertebrates, the cytosolic K^+ is 400 mM which is almost 20-fold higher compared to the blood, where it is only 20 mM. In vertebrates, K^+ concentration follows the same trend. In mammalian cell, the cytosolic concentration of K^+ is reported to be approximately 100 mM and up to 4 mM in blood (Lodish et al. 2000).

Potassium as an Osmoticum

When considering the osmotic balance of a cell, one will recall two ions, Na^+ and K^+. Nonetheless, both the ions hold equal importance in the context of osmolarity maintenance, but priorities may exist due to variable habitat of an organism (Fig. 1.1). K^+ is a well-known osmoticum at the cellular as well as physiological level ranging from bacteria to human (Wolfgang 1986). In bacteria, it is majorly required for the turgor maintenance of a single cell, and in complex eukaryotes, it is required by cells as well as tissues. Irrespective of the type of the cell, it serves a universal function of preservation of osmotic pressure inside the cells and helps them to cope up with hypo- or hyper-osmotic stress (Whatmore et al. 1990; Shabala and Lew 2002). In animals, mostly Na^+/K^+ exchangers are responsible for maintaining the osmotic potential, whereas in lower organisms and plants, K^+ channels and transporters do exist to contribute to osmotic balance within the organism. Studies based on the human erythrocytes over the years had made a clear indication of the link between osmotic stabilization in animal cells and active transport of Na^+ and K^+ at the cellular level (Armstrong 2003).

Potassium as a Signaling Molecule

The classic signaling molecules such as $3'-5'$ cyclic adenosine monophosphate (cAMP), diacylglycerol (DAG) and soluble messenger 1,4,5-inositol triphosphate (IP_3) are well worked out in the context of cell signaling. However, signaling molecule such as calcium (Ca^{2+}), a well-known second messenger, can act as nutrient at physiological level. Acting as a signal at the cellular level and on the top of it acting as a nutrient at physiological levels highlight the dual role of the molecule in regulating cellular physiology (Newton et al. 2016). There are likely chances that this concept may expand to the K^+. In bacteria, reports suggest K^+ is responsible for activating enzymes as well as transport systems resulting in adaptation to elevated osmolarity, hence acting as a cytoplasmic signaling molecule (Epstein, 2003). Although there are lack of evidence of K^+ as a signaling molecule in several organisms, there are speculations that it might serve as a signaling molecule in plants. Figure 1.1 summarizes the diverse roles played by K.

Importance of Potassium in Agriculture

Rapid expansion of world population in the past decades had a greater impact on agriculture, and if the trend persists, crop production with efficient use of ecosystem and other natural resources will be a 'must-to-do' requirement in the future. Increasing crop yield in resource-efficient system can be an 'impossible-to-achieve' task as the major hindrance in the path of crop yield is promptly striking abiotic and biotic stresses. Out of all the essential nutrients, the only nutrient which has a major impact on physiological processes vital to growth, yield, quality and stress resistance is K. In spite of presence of ample amount of K in the soil, only dissolved fraction of K is available to plants. Worldwide, large agricultural areas are reported to be K deficient including two-third of wheat belt of southern Australia and three-fourth of paddy soil of China (Mengel and Kirkby 2001). For crop maintenance, addition of K fertilizer is necessary. The texture of the soil is an important parameter for the low K availability as soils with low K status are often saline/acidic, sandy or waterlogged depending upon their K retaining capacity (Goulding and Loveland

1986). Existence of a precise dynamic equilibrium in the soil among all the available pools decides whether K needs to be fixed or released in the soil (Zorb et al. 2014). Soil fertility is drastically reduced upon lower K fertilizer application because of less abundance of plant-available K reserves in the soil. Even so this is not the only prime factor that determines the fixation as well as release of K in the soil. Several other factors such as plant-soil interactions and soil-microbial activities along with physical and chemical properties of the soil contribute to this process (Zorb et al. 2014). Due to the existence of strong binding forces between clay minerals and K, the process of K release is a relatively slow process compared to the K fixation (Öborn et al. 2008). Agriculturists may take advantage of the plant species effective in harbouring K-solubilizing bacteria and efficient K uptake from the soil, hence implying a control on K release from soil minerals (Zorb et al. 2014).

Potassium Availability in the Agricultural Soil: A Perpetual Need

In terms of nutritional status of K, crops are dependent on soil-available K that is controlled by several factors discussed above. A continuous need to boost the soil K availability has led to the introduction of few applications/approaches that may contribute significantly to the crop production. One of the widely used applications is the utilization of plant species or genotypes harbouring variable inherent capacity to use non-exchangeable K (NEK) sources as well as efficient absorption of K from the soil (Rengel 1993; Wang et al. 2011). This difference of utilizing NEK resources by various crop species has been reported in the past, where ryegrass and sugarbeet have been shown to be more efficient in mobilizing K compared to wheat and barley (Dessougi et al. 2002). The mobilization of NEKs is majorly mediated via the release of various organic acids such as citric acid, oxalic acid, tartaric acid and maleic acid by root exudates of crop plants (Fig. 1.2) (Kraffczyk et al. 1984; Chen et al. 2000). The organic acids either form metallo-organic complexes or enhance H^+/K^+ exchange aiding in weathering of soil minerals (Hinsinger and Jaillard 1993). Surprisingly, one of the reports suggested that enhanced release of K from clay minerals was also contributed by amino acid present in the root exudates (Rengel and Damon 2008). Root exudation mechanism gets activated when K level in the rhizosphere soil solution is below a threshold level of 10–20 µM (Hosseinpur et al. 2012; Schneider et al. 2013). For sustainable agriculture, unraveling the mechanism of K release from soil minerals is like a key to the door for developing new approaches.

Another alternative approach uses microorganisms such as *Pseudomonas* sp., *Burkholderia* sp., *Acidithiobacillus ferrooxidans*, *Bacillus mucilaginosus*, *Bacillus edaphicus*, *Bacillus megaterium*, etc. to increase the release of K from clay minerals (Fig. 1.2) (Fang et al. 2002). These microorganisms excrete organic acids as exudations that can perform K release in two different ways: either by directly dissolving

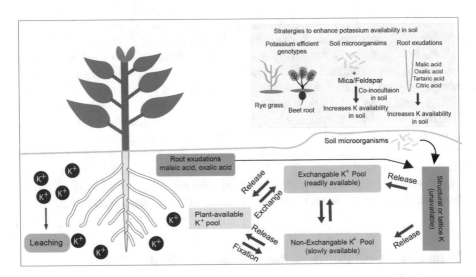

Fig. 1.2 Schematic representation of K$^+$ availability in the soil among the types of pools present in the soil, representing structural or lattice K, exchangeable K$^+$ pool, non-exchangeable K$^+$ pool and plant-available K$^+$ pool. Strategies to enhance K$^+$ acquisition from the soil have been also represented pictorially which includes usage of K-efficient genotypes, using K-releasing bacterial species and root exudations to increase K$^+$ availability in the soil. (Adapted and modified from Römheld and Kirkby (2010))

rock K or by chelating silicon ions from primary minerals to ensure K availability into the solution (Bennett et al. 1988; Basal and Biswas 2008). Concurrent application of K-solubilizing microorganisms and mica into the soil has been proved to be beneficial for the crops such as cotton, pepper, cucumber, oilseed rape and sudan grass (Fig. 1.2). Also, a large number of incubation trials performed using inoculated feldspar to the soil have shown 40–60% increase in K solubility and enhancement of plant K$^+$ uptake (Fang et al. 2002; Han and Lee 2005; Basal and Biswas 2008). Although the results of these incubation experiments seem to be promising and impeccable, as all the experiments were performed under stringent laboratory conditions, no strong conclusion can be drawn.

In order to support the rising demand of food production from the agricultural land, the current focus is to increase the K$^+$ use efficiency (KUE) (Srivastava et al. 2019). At the physiological level, KUE can be determined by monitoring two parameters. One of them is K uptake efficiency (KUpE) which is referred to as the ability of the roots to acquire K from the soil; another is K utilization efficiency (KUtE). KUtE is the ability of a plant to utilize acquired K from the soil to produce yield. Interestingly, the traits regulating KUpE and KUtE have been identified at whole plant level. Factors regulating KUpE are high root uptake capacity, early root vigour, high root-to-shoot ratios, high root length densities, proliferation of roots throughout the soil volume, high transpiration rates and exudation of organic compounds that are responsible for more non-exchangeable soil K. In a similar manner, KUtE is dependent on various factors such as effective K distribution within the plant, tolerance of the plant to low K concentrations, maintenance of optimal K concentrations in metabolically active cellular compartments, replacement of K in its non-specific roles, redistribution of K from senescent to younger tissues as well as maintenance of water relations and photosynthesis and a high harvest index (White 2013; Srivastava et al. 2019).

In order to reduce the demand of K fertilizers, another sustainable method has been proposed, utilizing K transporters as a means to improve plant KUE genetically. K transporters have been recognized and proposed to be the suitable targets for genetic manipulation in order to increase KUE which contributes to overall performance of the plant under extreme climate conditions (Wang and Wu 2015).

To further explore the potential of the above-mentioned approaches for agricultural production, field trials are an essential requirement to access their effect on both soil properties and crop growth (Zorb et al. 2014). It is very crucial for the agriculturists to understand the factors that regulate K release from soil non-exchangeable pool for better and optimized management practices. In addition to organic acids, K release from clay minerals also seems to be dependent on the H^+ concentration in soil solution. Therefore, optimization of soil pH might serve as a means of enhancing K release.

A broader role of K^+ which has been discussed in the chapter, starting from its importance as a nutrient to a signaling molecule, has been highlighted. Due to its importance among all forms of life, plant biologists are in a race to engineer crops for high K^+ uptake as well as K^+ utilizing efficiency.

References

Amtmann, A., Troufflard, S., & Armengaud, P. (2008). The effect of potassium nutrition on pest and disease resistance in plants. *Physiologia Plantarum, 133,* 682–691.

Armstrong, C. M. (2003). The Na/K pump, Cl ion, and osmotic stabilization of cells. *Proceedings of the National Academy of Sciences of the United States of America, 100,* 6257–6262.

Ashraf, K. U., Josts, I., Mosbahi, K., Kelly, S. M., Byron, O., Smith, B. O., & Walker, D. (2016). The potassium binding protein Kbp is a cytoplasmic potassium sensor. *Structure, 24,* 741–749.

Basal, B. B., & Biswas, D. R. (2008). Influence of potassium solubilizing microorganism (Bacillus mucilaginosus) and waste mica on potassium uptake dynamics by sudan grass (Sorghum vulgare Pers.) grown under two Alfisols | SpringerLink. *Plant and Soil, 317,* 235–255.

Bennett, P. C., Choi, W. J., & Rogers, J. R. (1988). Microbial destruction of feldspars. *Mineralogical Magazine, 8,* 149–150.

Chen, Y. X., Lin, Q., Lu, F., & He, Y. (2000). Study on detoxication of organic acid to raddish under the stress of Pb and Cd. *Acta Scientiae Circumstantiae, 20,* 467–472.

Dessougi, H. E., Claassen, N., & Steingrobe, B. (2002). Potassium efficiency mechanisms of wheat, barley, and sugar beet grown on a K fixing soil under controlled conditions. *Journal of Plant Nutrition and Soil Science, 165,* 732–737.

Dreyer, I., & Uozumi, N. (2011). Potassium channels in plant cells. *The FEBS Journal, 278,* 4293–4303.

Egilla, J. N., Davies, F. T., & Drew, M. C. (2001). Effect of potassium on drought resistance of Hibiscus rosa-sinensis cv. Leprechaun: Plant growth, leaf macro- and micronutrient content and root longevity. *Plant and Soil, 229,* 213–224.

Epstein, W. (2003). The roles and regulation of potassium in bacteria. *Progress in Nucleic Acid Research and Molecular Biology, 75*, 293–320.

Fang, S. X., Yan, H. L., & Yi, H. W. (2002). The conditions of releasing potassium by a silicate-dissolving bacterial strain NBT. *Agricultural Sciences in China, 1*, 662–666.

Goulding, K. W. T., & Loveland, P. J. (1986). The classification and mapping of potassium reserves in soils of England and Wales. *Journal of Soil Science, 37*, 555–565.

Han, H. S., & Lee, K. D. (2005). Phosphate and potassium solubilizing bacteria effect on mineral uptake, soil availability and growth of eggplant. *Research Journal of Agriculture and Biological Sciences, 1*, 176–180.

Hinsinger, P., & Jaillard, B. (1993). Root induced release of interlayer potassium and vermiculitization of phlogopite as related to potassium depletion in the rhizosphere of ryegrass. *Journal of Soil Science, 44*, 525–534.

Hosseinpur, A. R., Salehi, M. H., & Motaghian, H. R. (2012). Potassium release kinetics and its correlation with pinto bean (Phaseolus vulgaris) plant indices | Request PDF. *Plant Soil and Environment, 58*, 328–333.

Jeschke, W. D., Kirkby, E. A., Peuke, A. D., Pate, J. S., & Hartung, W. (1997). Effects of P deficiency on assimilation and transport of nitrate and phosphate in intact plants of castor bean (Ricinus communis L.). *Journal of Experimental Botany, 48*(1), 75–91.

Kraffczyk, I., Trolldenier, G., & Beringer, H. (1984). Soluble root exudates of maize: Influence of potassium supply and rhizosphere microorganisms. *Soil Biology and Biochemistry, 16*, 315–322.

Leigh, R. A., & Jones, R. G. W. (1984). A hypothesis relating critical potassium concentrations for growth to the distribution and functions of this ion in the plant cell. *New Phytologist, 97*(1), 1–13.

Lodish, H., Berk, A., Zipursky, S. L., Matsudaira, P., Baltimore, D., & Darnell, J. (2000). Intracellular ion environment and membrane electric potential. In S. Tenney (Ed.), *Molecular cell biology*. New York: W. H. Freeman.

Maroulis, S. L., Schofield, P. J., & Edwards, M. R. (2003). Osmoregulation in the parasitic protozoan Tritrichomonas foetus. *Applied and Environmental Microbiology, 69*, 4527–4533.

Marschner, P. (2012). *Mineral nutrition of higher plants*. London: Academic.

Mengel, K., & Kirkby, E. A. (2001). *Principles of plant nutrition*. Dordrecht: Springer.

Navarrete, C., Petrezselyova, S., Barreto, L., Martinez, J. L., Zahradka, J., Arino, J., Sychrova, H., & Ramos, J. (2010). Lack of main K+ uptake systems in Saccharomyces cerevisiae cells affects yeast performance in both potassium-sufficient and potassium-limiting conditions. *FEMS Yeast Research, 10*, 508–517.

Newton, A. C., Bootman, M. D., & Scott, J. D. (2016). Second messengers. *Cold Spring Harbor Perspectives in Biology, 8*, a005926.

Öborn, I., Rangel, A., Grant, C. A., Watson, C. A., & Edwards, A. C. (2008). Critical aspects of potassium management in agricultural systems. *Soil Use and Management, 21*, 102–112.

Rengel, Z. (1993). Mechanistic simulation models of nutrient uptake: A review | SpringerLink. *Plant and Soil, 152*, 161–173.

Rengel, Z., & Damon, P. M. (2008). Crops and genotypes differ in efficiency of potassium uptake and use. *Physiologia Plantarum, 133*, 624–636.

Römheld, V., & Kirkby, E. A. (2010). Research on potassium in agriculture: Needs and prospects | SpringerLink. *Plant and Soil, 335*, 155–180.

Rüdiger, H., & Ralf-Rainer, M. (1997). *Cell biology of metals and nutrients*. Berlin: Springer.

Schneider, A., Tesileanu, R., Charles, R., & Sinaj, S. (2013). Kinetics of soil potassium sorption–desorption and fixation. *Communications in Soil Science and Plant Analysis, 44*, 837–849.

Schroeder, D. (2019). Structure and weathering of potassium containing minerals. *IPI Research topics, 2013*, 43–63.

Shabala, S. N., & Lew, R. R. (2002). Turgor regulation in osmotically stressed Arabidopsis epidermal root cells. Direct support for the role of inorganic ion uptake as revealed by concurrent flux and cell turgor measurements. *Plant Physiology, 129*, 290–299.

Srivastava, A. K., Shankar, A., Chandran, A. K., Sharma, M., Jung, K. H., Suprasanna, P., & Pandey, G. K. (2019). Emerging concepts of potassium homeostasis in plants. *Journal of Experimental Botany, 71*(2), 608–619.

Umar, S., & Moinuddin. (2007). Genotypic differences in yield and quality of groundnut as affected by potassium nutrition under erratic rainfall conditions. *Journal of Plant Nutrition, 25*, 1549–1562.

Wang, Y., & Wu, W. H. (2015). Genetic approaches for improvement of the crop potassium acquisition and utilization efficiency. *Current Opinion in Plant Biology, 25*, 46–52.

Wang, H. Y., Shen, Q. H., Zhou, J. M., Wang, J., & Du, C. W. (2011). Plants use alternative strategies to utilize nonexchangeable potassium in minerals | SpringerLink. *Plant and Soil, 343*, 209–220.

Wedepohl, H. K. (1995). The composition of the continental crust. *Geochimica et Cosmochimica Acta, 59*, 1217–1232.

Whatmore, A. M., Chudek, J. A., & Reed, R. H. (1990). The effects of osmotic upshock on the intracellular solute pools of Bacillus subtilis. *Journal of General Microbiology, 136*, 2527–2535.

White, P. J. (2013). Improving potassium acquisition and utilisation by crop plants. *Journal of Plant Nutrition and Soil Science, 176*, 305–316.

Wolfgang, E. (1986). Osmoregulation by potassium transport in Escherichia coli. *FEMS Microbiology Reviews 2*, 73–78.

Zorb, C., Senbayram, M., & Peiter, E. (2014). Potassium in agriculture–status and perspectives. *Journal of Plant Physiology, 171*, 656 669.

Chapter 2
Potassium Homeostasis

Introduction

K is accumulated in various living cells where it is required for many physiological functions. If one looks carefully, one of the priority tasks of a cell is to maintain the K homeostasis, because of its vital role in cellular processes from bacteria to the most complex eukaryotes. In order to maintain K homeostasis, different organisms have evolved mechanisms that are quite variable from each other. In bacteria, numerous transport systems which differ in energy coupling, kinetics and their regulation are responsible for the tight regulation of K^+ in cytoplasmic pools. However, in eukaryotes where Na^+/K^+ exchangers are predominantly involved in K homeostasis in animals, in plants specialized channels and transporters are dedicated to maintain K homeostasis.

General Overview of Potassium Homeostasis: A Comparison Among Eukaryotes and Prokaryotes

As mentioned earlier, all organisms have evolved K homeostasis mechanisms that differ significantly from each other in terms of regulation and number of components required to maintain homeostasis. Overall K homeostasis is maintained by K uptake channels and transporters, K efflux carriers and Na^+/K^+ exchangers. The composition and abundance of these components of K homeostasis machinery varies from organism to organism depending on its complexity and external environmental conditions.

© The Author(s), under exclusive license to Springer Nature Switzerland AG 2020
G. K. Pandey, S. Mahiwal, *Role of Potassium in Plants*, SpringerBriefs in Plant
Science, https://doi.org/10.1007/978-3-030-45953-6_2

Prokaryotes

Controlled efflux and uptake of K⁺ is a critical determinant for growth and survival of bacteria due to their involvement in regulating pH and cell turgor (Roosild et al. 2010). In the model bacterium *Escherichia coli* (*E. coli*), high intracellular K⁺ is maintained by low-affinity K⁺ transporter (TrK) such as TrkG/H-TrkA and K⁺ uptake (KUP) and high-affinity transporters (KdpFABC). A well-known two-component system, the KdpD/kdpE system, is a major determinant of the K homeostasis where synthesis of high K⁺ uptake system is regulated by K⁺-dependent D (KdpD), a histidine kinase that acts as a K⁺ sensor along with the K⁺-dependent E (KdpE), a response regulator, in a phosphorylation-dependent manner (Fig. 2.1) (Laermann et al. 2013; Ashraf et al. 2016; Schramke et al. 2017). Though the mechanism seems to be very simple, as expected in a prokaryote, there are questions on the nature of the stimulus which is sensed by KdpD, and earlier the stimulus was thought to be 'turgor' (Laimins et al. 1981; Malli and Epstein 1998). Later on, the important piece of information further provided insights into this 'long-standing puzzle' and became the most accepted view of the mechanism (Schramke et al. 2017). Three chemical stimuli, i.e. extracellular and intracellular K⁺ concentration, ionic strength and ATP levels, are perceived by KdpD under limiting K⁺ condition, which leads to autophosphorylation of KdpD, and this further leads to phosphorylation of kdpE (Jung and Altendorf 1998; Heermann et al. 2000; Jung et al. 2000; Laermann et al. 2013; Heermann et al. 2014; Schramke et al. 2016). Phosphorylated

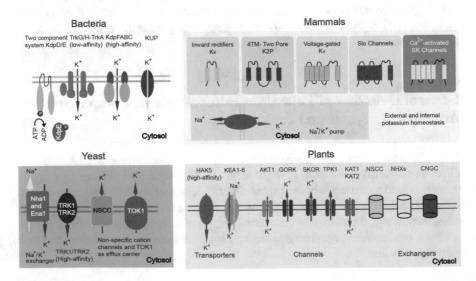

Fig. 2.1 Schematic representation of types of K⁺ uptake system in different organisms such as bacteria, yeast, mammals and plants. The number of K⁺ channels/transporters required to maintain homeostasis can be seen increasing as the complexity of the organism increases. Mammals and plants have more number of K⁺ transport systems compared to less-complex organisms such as bacteria and yeast

form of KdpE upon dimerization activates expression of high-affinity K^+ uptake system KdpFABC. However, KdpD also possesses phosphatase activity that can also dephosphorylate KdpE, finally suppressing the expression of KdpFABC in K^+ sufficiency conditions (Sugiura et al. 1992; Voelkner et al. 1993). Another low-affinity K^+ transporter TrkG/H-TrkA and KUP transporters are responsible for K^+ uptake in *E. coli* (Fig. 2.1) (Schlosser et al. 1995; Trchounian and Kobayashi 1999). The overall K homeostasis machinery has been pictorially represented in Fig. 2.1.

K^+ efflux systems (KefB and KefC) only function during metabolite detoxification in *E. coli*. Upon exposure to toxic electrophiles, including metabolic methylglyoxal, survival of *E. coli* is supported by Kef systems (Ferguson et al. 1993, 1995; Roosild et al. 2010). The KefB and KefC transport systems get activated by glutathione adducts and are inhibited by glutathione. Activation of Kef systems by glutathione S-conjugates integrates K^+ efflux with detoxification resulting in cytoplasmic acidification which protects against electrophile-mediated damage. KefB and KefC activity is present in a wide range of bacteria, including pathogenic enteric group such as *Pseudomonas* and *Staphylococcus aureus*, which suggests that regulation of K^+ can be an essential component of the survival strategy adopted by bacteria upon exposure to toxic compounds (Ferguson et al. 1993).

Yeast: The Eukaryotic Model System

In yeast, K^+ homeostasis is maintained by various components such as K^+ transporters (Trk1 and Trk2), acting as K^+ uptake system and outward rectifier K^+ channel (TOK1) acting as efflux carrier (Fig. 2.1) (Gaber et al. 1988; Ko et al. 1990; Ko and Gaber 1991). Yeast cells utilizes different strategies in order to maintain optimal intracellular K^+ concentrations along with maintenance of highly stable intracellular K^+/Na^+ ratio by utilizing K^+ transporters (Arino et al. 2010). One of the major strategies adopted by yeast cells is distinction between different cations, which favours the uptake of K^+ compared to Na^+, and this process heavily depends on the extracellular pH, where K^+ uptake is favoured at a lower pH (pH 4–6) (Armstrong and Rothstein 1964) (Rodríguez-Navarro and Ramos, 1984). The only other way for maintaining intracellular K^+ balance is to either extrude it or compartmentalize it. Efficient extrusion of excess toxic cations from cell or compartmentalization of cations into organelles such as vacuole is mediated by transport systems present at plasma membrane and organellar membranes (Arino et al. 2010). Remarkably known as high-affinity transport system, Trk1 sustains the capability of functioning as low-affinity or high-affinity transport system depending upon growth and K^+ status of the cell. Under standard growth conditions in wild-type strains, Trk1 acts as a major influx system, whereas upon starvation, Trk2 displays high-affinity transport in contrast to its poor expression under standard growth conditions (Gaber et al. 1988; Ko et al. 1990; Ramos et al. 1994; Arino et al. 2010). Earlier reports suggest that there are evidences of low-affinity transport in yeast cells, for which non-specific cation channel 1 (NSCC1) is suggested as a potential candidate (Bihler et al.

1998; Roberts et al. 1999; Giaever et al. 2002; Arino et al. 2010). Three different transport systems contribute to K⁺ efflux depending on the physiological conditions. Out of all three, two transporters, sodium transport ATPase1 (Ena1) and Na⁺/H⁺ antiporter (Nha1) are K⁺/Na⁺ efflux systems, whereas TOK1 is a K⁺-specific efflux system working as an outward-rectifying channel located at the yeast plasma membrane (Haro et al. 1991; Bertl et al. 1993; Banuelos et al. 2002; Ruiz and Arino 2007; Navarrete et al. 2010).

Potassium Homeostasis in Animals

Evolution of the first terrestrial cell is believed to have occurred in geothermal environments, possessing high K⁺/Na⁺ ratio along with relatively high concentrations of Zn^{2+}, Mn^{2+} and phosphorous compounds (Mulkidjanian et al. 2012). This is further supported by the fact that levels of K, phosphate and other transition metals are much higher in all the cells compared to modern marine resources (Mulkidjanian et al. 2012). Known for its role in regulating a broad array of physiological processes, K⁺ itself is under the tight regulation served by multiple mechanisms that collectively contributes to K homeostasis. In animals, narrow range maintenance of total K content in the body as well as plasma protein levels signifies 'bonafide K homeostasis' consisting of two synchronous processes: external K homeostasis and internal K homeostasis (Fig. 2.1) (Gumz et al. 2015). The overall process of external K homeostasis rectifies any K deficits, regulates renal K excretion to balance K intake and also disables extra-renal K loss (Elsevier, 31st December 2012; Greenlee et al. 2009). Further, three control systems entail external K homeostasis, out of which two systems can be categorized as 'reactive', whereas the third system is considered to be 'predictive'. Changes in the plasma K level are sensed by a negative-feedback system that regulates the K balance. K excretion increases in response to elevation in the plasma K level, which results in increasing K excretion, thus ultimately leading to a decline in K levels in plasma. Existence of a reactive feed-forward system, irrespective of the changes in the systemic plasma K level, is responsible for responding to K intake. Currently, the component mechanisms remain under study and are incompletely delineated (Greenlee et al. 2009; Gumz et al. 2015).

Internal K homeostasis is mainly concerned with the control of the asymmetric distribution of the total body K, which takes approximately 98% of intracellular fluid and only a small fraction (approx. 2%) of the extracellular fluid into account. This is brought about by the active cellular uptake through Na⁺/K⁺-ATPase and the passive K⁺ efflux from the cell which helps in achieving a balance of K⁺ within the cell (Gumz et al. 2015).

At the cellular level, complex array of K⁺ channels mediate K⁺ homeostasis in mammals. Structural classification of all K⁺ channels has led to their arrangement in three families. These are (i) inward rectifier (Kir) family, (ii) two-pore 4TM segment K⁺ channels and (iii) 6 TM segment K⁺ channels. Five major subfamilies have been defined, based on their structure, function and localization which include (i) inward rectifiers, Kir (7 subfamilies-15 genes); (ii) 4 transmembrane segments-2 pores, K2P (15 genes); (iii) voltage-gated, Kv (32 genes); (iv) the Slo family; and (v) Ca^{2+}-activated SK family, SKCa (Gonzalez et al. 2012; Huang and Jan 2014).

Potassium Homeostasis in Plants

Plants act as a dietary source of K for other living beings, and at the same time, K is an essential and abundant element required for regulating diverse physiological processes in plants. Hence, plants require an intricate system to maintain cellular K homeostasis. Differential expression of various channels and transporters located in the root, shoot and leaves based on the availability of K^+ in the soil contributes to K^+ balance at the cellular and tissue level as shown in Fig. 2.1. In *Arabidopsis* roots, K^+ transport is mediated majorly by *Arabidopsis* (shaker-type) K^+ channel (AKT1) (Rubio et al. 2008) and high-affinity K^+ transporters (HAK5) (Qi et al. 2008) with minor contributions of cyclic nucleotide-gated channels (CNGC) (Jha et al. 2016), guard cell outward-rectifying K^+ channel (GORK) (Hosy et al. 2003), tonoplast two-pore K^+ channel (TPK1) (Gobert et al. 2007) and Na^+/H^+ exchangers (NHXs) (Barragán et al. 2012). Non-selective cation channels (NSCC) (Tyerman 2002) and stellar outward-rectifying K^+ channels (SKOR) (Johansson et al. 2006) play a prime role in long-distance transport of K^+ from root cortex to xylem. In shoot, SKOR activity is a major determinant for maintaining shoot K^+ content, and vacuolar transport systems such as TPK, NHX and KUP/HAK are known for their minor contribution to K^+ homeostasis in shoot. In leaves, opening and closing of stomata is principally dependent on K^+ concentration of the guard cell. To balance the K^+ concentration inside the guard cell, the function of three channels has been studied extensively, two inward rectifiers, potassium channel in *Arabidopsis thaliana* (KAT1, KAT2) (Pilot et al. 2001), and one guard cell outward rectifier K^+ channel, GORK. However, another transport system such as NHX also has been implicated in this process (Shabala 2003). Recent studies have proven the K transport activity of K^+ efflux transporter (KEA1-6) suggesting their involvement in K^+ efflux (Tsujii et al. 2019). Though plant K uptake and transport systems have been studied extensively, still evidences are lacking for the regulatory mechanism that control the activity and spatiotemporal expression pattern of these channels/transporters. And this needs to be deciphered by functional characterization of modules controlling these components of 'K homeostasis machinery'.

A common feature to all organisms is presence of at least one K^+ channel in order to maintain K^+ homeostasis inside cells. Low-affinity as well as high-affinity uptake systems contribute to overall K^+ homeostasis among all forms of life. This can be an adaptive mechanism for ensuring K^+ availability from the surrounding environment.

References

Arino, J., Ramos, J., & Sychrova, H. (2010). Alkali metal cation transport and homeostasis in yeasts. *Microbiology and Molecular Biology Reviews, 74*, 95–120.

Armstrong, W. M., & Rothstein, A. (1964). Discrimination between alkali metal cations by yeast. I. Effect of PH on uptake. *The Journal of General Physiology, 48*, 61–71.

Ashraf, K. U., Josts, I., Mosbahi, K., Kelly, S. M., Byron, O., Smith, B. O., & Walker, D. (2016). The potassium binding protein Kbp is a cytoplasmic potassium sensor. *Structure, 24*, 741–749.

Banuelos, M. A., Ruiz, M. C., Jimenez, A., Souciet, J. L., Potier, S., & Ramos, J. (2002). Role of the Nha1 antiporter in regulating K(+) influx in Saccharomyces cerevisiae. *Yeast, 19*, 9–15.

Barragán, V., Leidi, E. O., Andrés, Z., Rubio, L., De Luca, A., Fernández, J. A., Cubero, B., & Pardo, J. M. (2012). Ion exchangers NHX1 and NHX2 mediate active potassium uptake into vacuoles to regulate cell turgor and stomatal function in Arabidopsis[W][OA]. *Plant Cell, 24*, 1127–1142.

Bertl, A., Slayman, C. L., & Gradmann, D. (1993). Gating and conductance in an outward-rectifying K+ channel from the plasma membrane of Saccharomyces cerevisiae. *The Journal of Membrane Biology, 132*, 183–199.

Bihler, H., Slayman, C. L., & Bertl, A. (1998). NSC1: A novel high-current inward rectifier for cations in the plasma membrane of Saccharomyces cerevisiae. *FEBS Letters, 432*, 59–64.

Elsevier. (2012, December 31). *Seldin and Giebisch's The Kidney*, 5th Edition.

Ferguson, G. P., Munro, A. W., Douglas, R. M., McLaggan, D., & Booth, I. R. (1993). Activation of potassium channels during metabolite detoxification in Escherichia coli. *Molecular Microbiology, 9*, 1297–1303.

Ferguson, G. P., McLaggan, D., & Booth, I. R. (1995). Potassium channel activation by glutathione-S-conjugates in Escherichia coli: Protection against methylglyoxal is mediated by cytoplasmic acidification. *Molecular Microbiology, 17*, 1025–1033.

Gaber, R. F., Styles, C. A., & Fink, G. R. (1988). TRK1 encodes a plasma membrane protein required for high-affinity potassium transport in Saccharomyces cerevisiae. *Molecular and Cellular Biology, 8*, 2848–2859.

Giaever, G., Chu, A. M., Ni, L., Connelly, C., Riles, L., Veronneau, S., Dow, S., Lucau-Danila, A., Anderson, K., Andre, B., Arkin, A. P., Astromoff, A., El-Bakkoury, M., Bangham, R., Benito, R., Brachat, S., Campanaro, S., Curtiss, M., Davis, K., Deutschbauer, A., Entian, K. D., Flaherty, P., Foury, F., Garfinkel, D. J., Gerstein, M., Gotte, D., Guldener, U., Hegemann, J. H., Hempel, S., Herman, Z., Jaramillo, D. F., Kelly, D. E., Kelly, S. L., Kotter, P., LaBonte, D., Lamb, D. C., Lan, N., Liang, H., Liao, H., Liu, L., Luo, C., Lussier, M., Mao, R., Menard, P., Ooi, S. L., Revuelta, J. L., Roberts, C. J., Rose, M., Ross-Macdonald, P., Scherens, B., Schimmack, G., Shafer, B., Shoemaker, D. D., Sookhai-Mahadeo, S., Storms, R. K., Strathern, J. N., Valle, G., Voet, M., Volckaert, G., Wang, C. Y., Ward, T. R., Wilhelmy, J., Winzeler, E. A., Yang, Y., Yen, G., Youngman, E., Yu, K., Bussey, H., Boeke, J. D., Snyder, M., Philippsen, P., Davis, R. W., & Johnston, M. (2002). Functional profiling of the Saccharomyces cerevisiae genome. *Nature, 418*, 387–391.

Gobert, A., Isayenkov, S., Voelker, C., Czempinski, K., & Maathuis, F. J. (2007). The two-pore channel TPK1 gene encodes the vacuolar K+ conductance and plays a role in K+ homeostasis. *Proceedings of the National Academy of Sciences of the United States of America, 104*, 10726–10731.

Gonzalez, C., Baez-Nieto, D., Valencia, I., Oyarzun, I., Rojas, P., Naranjo, D., & Latorre, R. (2012). K(+) channels: Function-structural overview. *Comprehensive Physiology, 2*, 2087–2149.

Greenlee, M., Wingo, C. S., McDonough, A. A., Youn, J. H., & Kone, B. C. (2009). Narrative review: Evolving concepts in potassium homeostasis and hypokalemia. *Annals of Internal Medicine, 150*, 619–625.

Gumz, M. L., Rabinowitz, L., & Wingo, C. S. (2015). An integrated view of potassium homeostasis. *The New England Journal of Medicine, 373*, 60–72.

Haro, R., Garciadeblas, B., & Rodriguez-Navarro, A. (1991). A novel P-type ATPase from yeast involved in sodium transport. *FEBS Letters, 291*, 189–191.

Heermann, R., Altendorf, K., & Jung, K. (2000). The hydrophilic N-terminal domain complements the membrane-anchored C-terminal domain of the sensor kinase KdpD of Escherichia coli. *The Journal of Biological Chemistry, 275*, 17080–17085.

Heermann, R., Zigann, K., Gayer, S., Rodriguez-Fernandez, M., Banga, J. R., Kremling, A., & Jung, K. (2014). Dynamics of an interactive network composed of a bacterial two-component system, a transporter and K+ as mediator. *PLoS One, 9*, e89671.

Hosy, E., Vavasseur, A., Mouline, K., Dreyer, I., Gaymard, F., Poree, F., Boucherez, J., Lebaudy, A., Bouchez, D., Very, A. A., Simonneau, T., Thibaud, J. B., & Sentenac, H. (2003). The Arabidopsis outward K+ channel GORK is involved in regulation of stomatal movements and plant transpiration. *Proceedings of the National Academy of Sciences of the United States of America, 100*, 5549–5554.

Huang, X., & Jan, L. Y. (2014). Targeting potassium channels in cancer. *The Journal of Cell Biology, 206*, 151–162.

Jha, S. K., Sharma, M., & Pandey, G. K. (2016). Role of cyclic nucleotide gated channels in stress management in plants. *Current Genomics, 17*, 315–329.

Johansson, I., Wulfetange, K., Poree, F., Michard, E., Gajdanowicz, P., Lacombe, B., Sentenac, H., Thibaud, J. B., Mueller-Roeber, B., Blatt, M. R., & Dreyer, I. (2006). External K+ modulates the activity of the Arabidopsis potassium channel SKOR via an unusual mechanism. *The Plant Journal, 46*, 269–281.

Jung, K., & Altendorf, K. (1998). Truncation of amino acids 12-128 causes deregulation of the phosphatase activity of the sensor kinase KdpD of Escherichia coli. *The Journal of Biological Chemistry, 273*, 17406–17410.

Jung, K., Veen, M., & Altendorf, K. (2000). K+ and ionic strength directly influence the autophosphorylation activity of the putative turgor sensor KdpD of Escherichia coli. *The Journal of Biological Chemistry, 275*, 40142–40147.

Ko, C. H., & Gaber, R. F. (1991). TRK1 and TRK2 encode structurally related K+ transporters in Saccharomyces cerevisiae. *Molecular and Cellular Biology, 11*, 4266–4273.

Ko, C. H., Buckley, A. M., & Gaber, R. F. (1990). TRK2 is required for low affinity K+ transport in Saccharomyces cerevisiae. *Genetics, 125*, 305–312.

Laermann, V., Cudic, E., Kipschull, K., Zimmann, P., & Altendorf, K. (2013). The sensor kinase KdpD of Escherichia coli senses external K+. *Molecular Microbiology, 88*, 1194–1204.

Laimins, L. A., Rhoads, D. B., & Epstein, W. (1981). Osmotic control of kdp operon expression in Escherichia coli. *Proceedings of the National Academy of Sciences of the United States of America, 78*, 464–468.

Malli, R., & Epstein, W. (1998). Expression of the Kdp ATPase is consistent with regulation by turgor pressure. *Journal of Bacteriology, 180*, 5102–5108.

Mulkidjanian, A. Y., Bychkov, A. Y., Dibrova, D. V., Galperin, M. Y., & Koonin, E. V. (2012). Origin of first cells at terrestrial, anoxic geothermal fields. *Proceedings of the National Academy of Sciences of the United States of America, 109*, E821–E830.

Navarrete, C., Petrezselyova, S., Barreto, L., Martinez, J. L., Zahradka, J., Arino, J., Sychrova, H., & Ramos, J. (2010). Lack of main K+ uptake systems in Saccharomyces cerevisiae cells affects yeast performance in both potassium-sufficient and potassium-limiting conditions. *FEMS Yeast Research, 10*, 508–517.

Pilot, G., Lacombe, B., Gaymard, F., Cherel, I., Boucherez, J., Thibaud, J. B., & Sentenac, H. (2001). Guard cell inward K+ channel activity in arabidopsis involves expression of the twin channel subunits KAT1 and KAT2. *The Journal of Biological Chemistry, 276*, 3215–3221.

Qi, Z., Hampton, C. R., Shin, R., Barkla, B. J., White, P. J., & Schachtman, D. P. (2008). The high affinity K+ transporter AtHAK5 plays a physiological role in planta at very low K+ concentrations and provides a caesium uptake pathway in Arabidopsis. *Journal of Experimental Botany, 59*, 595–607.

Ramos, J., Alijo, R., Haro, R., & Rodriguez-Navarro, A. (1994). TRK2 is not a low-affinity potassium transporter in Saccharomyces cerevisiae. *Journal of Bacteriology, 176*, 249–252.

Roberts, S. K., Fischer, M., Dixon, G. K., & Sanders, D. (1999). Divalent cation block of inward currents and low-affinity K+ uptake in Saccharomyces cerevisiae. *Journal of Bacteriology, 181*, 291–297.

Rodríguez-Navarro, A., & Ramos, J. (1984). Dual system for potassium transport in Saccharomyces cerevisiae. *Journal of Bacteriology, 159*, 940–945.

Roosild, T. P., Castronovo, S., Healy, J., Miller, S., Pliotas, C., Rasmussen, T., Bartlett, W., Conway, S. J., & Booth, I. R. (2010). Mechanism of ligand-gated potassium efflux in bacterial pathogens. *Proceedings of the National Academy of Sciences of the United States of America, 107*, 19784–19789.

Rubio, F., Nieves-Cordones, M., Aleman, F., & Martinez, V. (2008). Relative contribution of AtHAK5 and AtAKT1 to K+ uptake in the high-affinity range of concentrations. *Physiologia Plantarum, 134*, 598–608.

Ruiz, A., & Arino, J. (2007). Function and regulation of the Saccharomyces cerevisiae ENA sodium ATPase system. *Eukaryotic Cell, 6*, 2175–2183.

Schlosser, A., Meldorf, M., Stumpe, S., Bakker, E. P., & Epstein, W. (1995). TrkH and its homolog, TrkG, determine the specificity and kinetics of cation transport by the Trk system of Escherichia coli. *Journal of Bacteriology, 177*, 1908–1910.

Schramke, H., Tostevin, F., Heermann, R., Gerland, U., & Jung, K. (2016). A dual-sensing receptor confers robust cellular homeostasis. *Cell Reports, 16*, 213–221.

Schramke, H., Laermann, V., Tegetmeyer, H. E., Brachmann, A., Jung, K., & Altendorf, K. (2017). Revisiting regulation of potassium homeostasis in Escherichia coli: The connection to phosphate limitation. *Microbiology, 6*.

Shabala, S. (2003). Regulation of potassium transport in leaves: From molecular to tissue level. *Annals of Botany, 92*, 627–634.

Sugiura, A., Nakashima, K., Tanaka, K., & Mizuno, T. (1992). Clarification of the structural and functional features of the osmoregulated kdp operon of Escherichia coli. *Molecular Microbiology, 6*, 1769–1776.

Trchounian, A., & Kobayashi, H. (1999). Kup is the major K+ uptake system in Escherichia coli upon hyper-osmotic stress at a low pH. *FEBS Letters, 447*, 144–148.

Tsujii, M., Kera, K., Hamamoto, S., Kuromori, T., Shikanai, T., & Uozumi, N. (2019). Evidence for potassium transport activity of Arabidopsis KEA1-KEA6. *Scientific Reports, 9*, 10040.

Tyerman, S. D. (2002). Nonselective cation channels. Multiple functions and commonalities. *Plant Physiology, 128*, 327–328.

Voelkner, P., Puppe, W., & Altendorf, K. (1993). Characterization of the KdpD protein, the sensor kinase of the K(+)-translocating Kdp system of Escherichia coli. *European Journal of Biochemistry, 217*, 1019–1026.

Chapter 3
Potassium Uptake and Transport System in Plant

Introduction

In plants, the only means of K^+ uptake from the soil is mediated by the root system where epidermal and cortical cells are responsible for most of the K^+ acquisition (Ashley et al. 2006). Membrane-localized transport proteins via symplast facilitate transport of K^+ inside the plant cell, as lipid bilayer of the cell membrane is impermeable to ions. Process of K^+ uptake follows biphasic kinetics where high-affinity K^+ transport (an active process) is mediated by H^+/K^+ symporters and low-affinity K^+ transport (a passive process) is mediated by channels (Alberts et al. 2002). High-affinity transport system shows saturating kinetics at K_m between 10 and 40 μM which is highly selective for K^+ over Na^+, whereas low-affinity transport system shows saturation at K_m near about 10 mM range exhibiting weak selectivity for K^+ over other alkali metals. The journey of K^+ from soil to its delivery to the respective plant tissue is possible because of concerted action of various channels and transporters.

Potassium Channels and Transporter Families in Plant

In the post-genomics era, an outburst of sequence information enabled identification of various K^+-permeable channels and transporters in *Arabidopsis thaliana* genome. Fifteen candidate genes have been identified encoding K^+ channels and three families of genes encoding transporters (Dreyer et al. 2011). Molecular characterization of these channels and transporter has led to their identification as low-affinity and high-affinity uptake components. Evidences suggest that transporters are a part of high-affinity uptake system (HATS), whereas channels contribute to low-affinity uptake systems (LATS). However, even if they have been studied

© The Author(s), under exclusive license to Springer Nature Switzerland AG 2020
G. K. Pandey, S. Mahiwal, *Role of Potassium in Plants*, SpringerBriefs in Plant Science, https://doi.org/10.1007/978-3-030-45953-6_3

extensively well, no clear distinctions can be made for the functionality of channels and transporters in the context of HATS and LATS (Dreyer and Uozumi 2011). These channels and transporters are further subdivided into different families.

Potassium Transporters

K+ transporters are mainly responsible for K+ transport and ion homeostasis in plants. Many of them have been recently identified and are well known to play roles in diverse physiological responses. Three major families of transporters majorly contribute to K homeostasis in plants: HAK/KT/KUP family, TRK/HKT family and cation/proton antiporter (CPA) families (Fig. 3.1). The HAK/KT/KUP family consists of 13 members, and CPA family (further classified into CPA1 and CPA2) consists of 42 members, whereas only 1 member has been identified as belonging to TRK/HKT family in *A. thaliana* (Grabov 2007). Structural variation does exist among all these transporters, each family exhibiting a distinct feature, which may attribute to their versatile functioning in the plant. The structural scaffold of transporters of HAK/KT/KUP family in rice and *Arabidopsis* comprised of 10–14 transmembrane domains and a cytoplasmic C-terminal tail (Gierth and Maser 2007). TRK/HKT family is thought to evolve from simple K+ channels. Characterized by pore loop in between two transmembrane domains, this family carries the

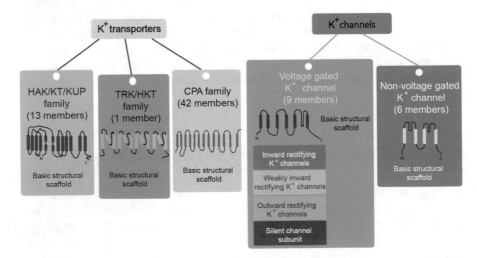

Fig. 3.1 Functional classification of K+ transporters and channels in plants. Transporters consist of 3 families: HAK family consisting of 13 members, TRK family consisting of only 1 member and CPA family having 42 members. Channels are subdivided into two categories based on their function into voltage-gated and non-voltage-gated. Voltage-gated K+ channels, consisting of nine members, are further classified into inward-rectifying K+ channels, weakly inward-rectifying K+ channels, outward-rectifying K+ channels and silent K+ channels. Non-voltage-gated K+ channels consist of six members

membrane-pore-membrane ('MPM') motif as building blocks, which are again structurally similar to some K^+ channels (Kato et al. 2001; Gierth and Maser, 2007).

Potassium Channels

Based on their gating activity, K^+ channels in plants are further divided into two classes: voltage-gated K^+ channels and non-voltage-gated K^+ channels.

Voltage-Gated K^+ Channels Voltage-gated K^+ channels are tetrameric proteins containing four α- subunits (Dreyer et al. 1997). These four subunits can be identical or different from each other; four identical subunits form homodimer, whereas different subunit compositions form heterodimer. Upon aggregation, the four subunits contribute to the central pore (P), which is a major determinant of the permeation properties of the channel, thus participating in the process. Nine members of this family have been identified in *Arabidopsis* so far (Lebaudy et al. 2008; Gajdanowicz et al. 2009).

Voltage-gated K^+ channels are diversified into four functional categories: inward-rectifying K^+ (K_{in}) channels responsible for K^+ uptake, silent channel (K_{silent}) subunits that regulate or modify properties of K_{in} channels, weakly inward-rectifying K^+ channels (K_{weak}) having special gating properties and outward-rectifying K^+ channels (K_{out}) mainly responsible for K^+ release (Fig. 3.1) (Pilot et al. 2003a; Poree et al. 2005; Dreyer et al. 2011). In *Arabidopsis*, *At*KAT1, *At*KAT2, *At*AKT1, *At*AKT5 and *At*SPIK are the members of K_{in} subgroup, *At*SKOR and *At*GORK belong to K_{out} subgroup, and *At*KC1 and *At*AKT2 are affiliated to K_{silent} and K_{weak}, respectively (Dreyer and Uozumi 2011).

Non-voltage-Gated K^+ Channels/Voltage-Independent K^+ Channels Non-voltage-gated K^+ channels are characterized by four transmembrane domains (TM domain) and two pore loops per subunit, hence resembling a duplicated K_{ir}-like structure (mammalian tandem P-domain channels) (Voelker et al. 2010). A total of six members, five tandem pore channels (TPKs) and a single subunit channel known as KCO3, contribute to the group of non-voltage-gated K^+ channels/voltage-independent K^+ channels in plants (Fig. 3.1) (Dreyer et al. 2011). Collectively, five TPKs (TPK1–TPK5) and a single KCO3 represent the AtTPK/KCO family in *A. thaliana* (Voelker et al. 2006). Several studies propose that in order to form a functional channel TPKs tend to make dimers consisting of identical subunits (Voelker et al. 2006).

Regulatory Mechanism of Potassium Channels and Transporters

Plant K⁺ transport systems (channels/transporters) are diversified and are mostly encoded by large gene families. Linking this diversity to the regulation of K⁺ transport in the context of overall plant physiology might be an enticing prospect for now, but dependency of various cell functions on the regulated K⁺ fluxes cannot be ignored (Very and Sentenac 2003). A deeper insight into molecular mechanisms regulating K⁺ transport systems in plants is the need of the hour to understand the fine-tuning of K⁺ balance that exists naturally in the plant cells.

Transcriptional and Post-translational Regulation of Potassium Transporters and Channels

In recent years, floods of information have infiltrated the current understanding of regulome (set of regulatory components in a cell) of K⁺ transport systems. Out of all, the regulatory aspects of a few channels and transporters have been worked out extensively such as AKT1, HAK5 and KUP4 in *A. thaliana* and other plant species. Few observations have established the fact that two major K⁺ uptake systems in plants present in roots *At*AKT1 and *At*HAK5 are regulated transcriptionally as well as by post-translational mechanisms. Experimental evidences suggesting transcriptional regulation of these two candidate genes suggest that the expression of *AKT1* is known to be downregulated upon treatment with hormone such as cytokinins and auxin (Pilot et al. 2003b). However, the expression of *HAK5* is regulated by certain factors such as nutrient deficiency, membrane potential, metabolites, ROS production, phytohormone and transcription factors (Shin and Schachtman 2004; Jung et al. 2009; Ragel et al. 2019). One of the major highlights of these studies is that K⁺ starvation leads to mRNA accumulation of *At*HAK5, which reduces significantly upon K⁺ resupply (Gierth et al. 2005). Experimental evidences also indicate that phosphate (P_i) deprivation also induces *At*HAK5 expression (Shin et al. 2005). Another study reveals that in response to K⁺ deficiency, RDH2/RbohC, an NADPH oxidase, produces H_2O_2 that in turn regulates HAK5 expression (Shin and Schachtman 2004). Metabolites such as sucrose contribute to increased expression of HAK5 in the dark in comparison to light, indicating photosynthesis-mediated regulation of HAK5 expression (Lejay et al. 2008). In roots of *A. thaliana*, salinity-induced membrane hyperpolarization due to massive change in membrane potential is responsible for inducing expression of *HAK5*- and *HAK1*-type transporters (Aleman et al. 2011). Phytohormone ethylene is also known to induce the expression of *HAK5* in *Arabidopsis* and tomato (Jung et al. 2009). Another study provided brief insights into transcription factor-mediated regulation of *HAK5*, and it is the only evidence so far describing negative regulation of the *HAK5* transcription by transcription factor auxin response factor 2 (ARF2) (Fig. 3.2) (Zhao et al. 2016).

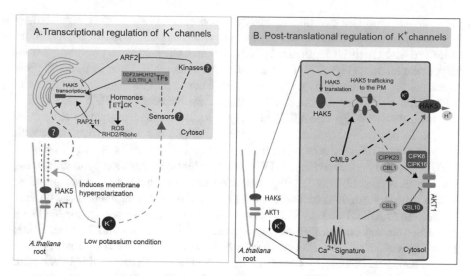

Fig. 3.2 Schematic representation of regulatory mechanisms controlling the activity of K$^+$ channels and transporters. (**a**) Transcriptional regulation occurs via various mechanisms such as membrane hyperpolarization, hormones and transcription factors in low K$^+$ conditions. (**b**) Post-translational control is mediated by CBL-*CIPK* complex and CML9. HAK5 activity is regulated at transcriptional as well as post-translational level by the above-mentioned factors, whereas AKT1 is known to be regulated at post-translational level. (Adapted and modified from Ragel et al. 2019)

Recently few reports have successfully shown that low K$^+$ concentration is a prerequisite for high-affinity K$^+$ uptake by HAK5, which further provides indication about presence of a post-translational mechanism in planta (Ragel et al. 2019). It was soon established that *At*HAK5 activity is modulated by a complex of calcineurin B-like proteins (CBL)/calcineurin B-like protein kinases (CIPKs) encompassing CBL1/9 and CIPK23 (Ragel et al. 2015). Complementation experiments performed in yeast suggest that other *Arabidopsis* CBLs (*At*CBL8/9/10) are also known to interact with CIPK23 resulting in the activation of *At*HAK5 (Fig. 3.2) (Ragel et al. 2019). Behera et al. (2017) precisely explained the involvement of CBL and CIPK complexes in this process. As plants are subjected to K$^+$ deficiency, a specific, biphasic calcium (Ca^{2+}) signature is generated that is sensed by CBL, probably leading to the activation of AtHAK5 through CBL-CIPK complex (Ragel et al. 2015). However, the differential detection of *At*HAK5 protein in endoplasmic reticulum and plasma membrane in K$^+$ sufficiency and K$^+$ deficiency conditions indicates the possibility of CBL1/CIPK23-mediated trafficking of HAK5 transporter to the plasma membrane (Qi et al. 2008). A new concept that has come to the limelight recently is the existence of second Ca^{2+}-dependent pathway to regulate *At*HAK5 activity. It has been proposed that INTEGRIN LINKED KINASE1 (ILK1) and calmodulin-like protein 9 (CML9) interact with HAK5 resulting in its accumulation on the plasma membrane. However, it is still a tempting conjecture that ILK1 may regulate HAK5 by phosphorylation since there is not much evidence for in vitro phosphorylation (Fig. 3.2) (Brauer et al. 2016).

Another important component of K^+ response, AKT1, is mainly regulated by post-translational mechanisms as evidenced by various studies. In low K^+ condition, AKT1 is known to be regulated by CBL1/9 and CIPK23 in a phosphorylation-dependent manner (Fig. 3.2) (Li et al. 2006; Xu et al. 2006; Luan et al. 2009). Few other reports have emerged with the evidence that AKT1 is regulated by multiple CBL-CIPK pairs where, besides CIPK23, the involvement of CIPK6 and CIPK16 has been shown to regulate AKT1 activity in a CBL-dependent manner (Lan et al. 2011). To positively regulate AKT1 activity, CIPKs are known to interact with ankyrin repeat domain present within the C-terminal region of AKT1 (Lee et al. 2007). It is also speculated that AKT1 activity might be controlled upon interacting with other alpha-subunits leading to the formation of heterotetrameric proteins, which may perform diverse functions. An example of this phenomenon can be seen in the case of *At*KC1 and *At*AKT1 where *At*KC1 regulates the activity of AKT1 by forming a ternary complex along with syntaxin SYP121 (Very and Sentenac 2003; Lebaudy et al. 2007). In addition to its regulation by CBL1/9-CIPK23 pair, AKT1 is also regulated directly by CBL10, independent of any CIPK. In this interaction, CBL10 competes with CIPK23 for binding with AKT1 and thus represses its activity (Ren et al. 2013).

Phosphorylation (or dephosphorylation) represents the prominent means of regulation whenever a kinase is involved. Interestingly, K^+ channel AKT2 is regulated by translocation from endoplasmic reticulum to plasma membrane by CBL4-CIPK6 pair where phosphorylation is not involved. Though not yet completely resolved, AKT2 activation by CBL4-CIPK6 has been proposed to be facilitated by forming a scaffold around the channel (Held et al. 2011).

Subcellular Localization

The structural 'prototype' of K^+ channels suggests their conservation among kingdoms; also they are present in every organism. In the context of subcellular localization, K^+ channels show harmony, as most of them are localized to plasma membrane and endomembranes. The distribution of the channels at the plasma membrane is not random; in fact, their distribution is affected by their interaction with the scaffolding proteins. For example, in human neuronal cells, K^+ channels are known to cluster at specific regions on the plasma membrane (O'Connell et al. 2006).

In plants, voltage-gated channels were usually found at the plasma membrane with an exception of TPK4 and TPK3, whereas non-voltage-gated channels were found to be located on endomembranes of different organelles. Among all the voltage-gated ion channels, localization experiments of TPK suggest TPK4 localization at disparate locations. Transient expression studies in *A. thaliana* protoplast demonstrate vacuolar localization for TPK4 (Voelker et al. 2006), whereas transient experiment studies in onion peel cells reveal that major fraction of the TPK4 is present in the endoplasmic reticulum (ER) with partial localization at the plasma membrane (Becker et al. 2004; Dunkel et al. 2008). Both the experiments were performed using TPK4:GFP fusion constructs so the possibility of differential targeting due to the 'tag' used has to be negated. Presence of TPK4 in the ER could be possibly because of its retention in the ER or the typical dual localization behaviour (Sharma et al. 2013). TPK3 also display dual localization behaviour, being present in the tonoplast as well as in thylakoid membrane (Dunkel et al. 2008; Sharma et al. 2013). In order to unravel whether any 'guiding sequence' targets TPKs to the membrane, chimeras of TPK4 and TPK1 were generated. Analysis of three diacidic motifs in the C-terminal showed that D296G/E298G mutations in one of these motifs lead to the ER retention of TPK4/TPK1 chimera. These observations suggest export of TPK1 from ER is dependent on the presence of this diacidic motif (Dunkel et al 2008; Voelker et al. 2010). Further observations suggested a Golgi-dependent pathway is responsible for the transport of TPK channel proteins to the vacuolar membrane, where, after its successful departure from the ER, Golgi apparatus is the first compartment crossed by this protein (Dunkel et al. 2008).

K^+ transporters belonging to the HAK and HKT transporter family are universally present on the plasma membrane in *Arabidopsis* and rice. Some of the KUP transporters (belonging to HAK family) have been shown to be localized at the tonoplast membrane in *Arabidopsis*. Anomalous localization of only one transporter, *OsHAK10*, has been found at tonoplast membrane in rice (Ragel et al. 2019). Three members of KEA family antiporters (having K^+ transport ability) have been shown to localize in the chloroplast. KEA1 and KEA2 localize to the inner envelope of chloroplast, whereas KEA3 has been reported to be targeted to the thylakoid membrane as well as Golgi apparatus (Zheng et al. 2013; Kunz et al. 2014). KEA4/5/6 have been shown to localize to Golgi apparatus (Wang et al. 2019).

Plants contain a number of channels and transporters located in various parts of the plants based on the K^+ requirement of the plant tissues. These channels and transporters are responsible for low-affinity as well as high-affinity uptake depending on the K^+ availability in the soil. K^+ is distributed in the whole plant body as per the activity of these transport systems. In microenvironment of a cell, they are usually present on the plasma membrane and endomembrane with some exceptions that are briefly discussed in the chapter.

References

Alberts, B., Johnson, A., Lewis, J., Raff, M., Roberts, K., & Walter, P. (2002). Principles of membrane transport. In *Molecular biology of the cell* (4th ed.). New York: Garland Science.

Aleman, F., Nieves-Cordones, M., Martinez, V., & Rubio, F. (2011). Root K(+) acquisition in plants: The Arabidopsis thaliana model. *Plant & Cell Physiology, 52*, 1603–1612.

Ashley, M. K., Grant, M., & Grabov, A. (2006). Plant responses to potassium deficiencies: A role for potassium transport proteins. *Journal of Experimental Botany, 57*, 425–436.

Becker, D., Geiger, D., Dunkel, M., Roller, A., Bertl, A., Latz, A., Carpaneto, A., Dietrich, P., Roelfsema, M. R., Voelker, C., Schmidt, D., Mueller-Roeber, B., Czempinski, K., & Hedrich, R. (2004). AtTPK4, an Arabidopsis tandem-pore K+ channel, poised to control the pollen membrane voltage in a pH- and Ca2+−dependent manner. *Proceedings of the National Academy of Sciences of the United States of America, 101*, 15621–15626.

Behera, S., Long, Y., Schmitz-Thom, I., Wang, X. P., Zhang, C., Li, H., Steinhorst, L., Manishankar, P., Ren, X. L., Offenborn, J. N., Wu, W. H., Kudla, J., & Wang, Y. (2017). Two spatially and temporally distinct Ca(2+) signals convey Arabidopsis thaliana responses to K(+) deficiency. *The New Phytologist, 213*, 739–750.

Brauer, E. K., Ahsan, N., Dale, R., Kato, N., Coluccio, A. E., Pineros, M. A., Kochian, L. V., Thelen, J. J., & Popescu, S. C. (2016). The Raf-like kinase ILK1 and the high affinity K+ transporter HAK5 are required for innate immunity and abiotic stress response. *Plant Physiology, 171*, 1470–1484.

Dreyer, I., Antunes, S., Hoshi, T., Müller-Röber, B., Palme, K., Pongs, O., Reintanz, B., Hedrich, R. (1997) Plant K+ channel alpha-subunits assemble indiscriminately. *Biophysical Journal, 72*(5), 2143–2150.

Dreyer, I., & Uozumi, N. (2011). Potassium channels in plant cells. *The FEBS Journal, 278*, 4293–4303.

Dreyer, I., Plant Biophysics and Heisenberg Group of Biophysics and Molecular Plant Biology, C.d.B.y.G.d.P, Universidad Politécnica de Madrid, Spain, Uozumi, N., Department of Biomolecular Engineering, G.S.o.E, & Tohoku University, Japan. (2011). Potassium channels in plant cells. *The FEBS Journal, 278*, 4293–4303.

Dunkel, M., Latz, A., Schumacher, K., Muller, T., Becker, D., & Hedrich, R. (2008). Targeting of vacuolar membrane localized members of the TPK channel family. *Molecular Plant, 1*, 938–949.

Gajdanowicz, P., Garcia-Mata, C., Gonzalez, W., Morales-Navarro, S. E., Sharma, T., Gonzalez-Nilo, F. D., Gutowicz, J., Mueller-Roeber, B., Blatt, M. R., & Dreyer, I. (2009). Distinct roles of the last transmembrane domain in controlling Arabidopsis K+ channel activity. *The New Phytologist, 182*, 380–391.

Gierth, M., & Maser, P. (2007). Potassium transporters in plants – involvement in K+ acquisition, redistribution and homeostasis. *FEBS Letters, 581*, 2348–2356.

Gierth, M., Mäser, P., & Schroeder, J. I. (2005). The potassium transporter AtHAK5 functions in K+ deprivation-induced high-affinity K+ uptake and AKT1 K+ channel contribution to K+ uptake kinetics in Arabidopsis roots1[w]. *Plant Physiology, 137*, 1105–1114.

Grabov, A. (2007). Plant KT/KUP/HAK potassium transporters: Single family – Multiple functions. *Annals of Botany, 99*, 1035–1041.

Held, K., Pascaud, F., Eckert, C., Gajdanowicz, P., Hashimoto, K., Corratgé-Faillie, C., Offenborn, J. N., Lacombe, B., Dreyer, I., Thibaud, J. B., & Kudla, J. (2011). Calcium-dependent modulation and plasma membrane targeting of the AKT2 potassium channel by the CBL4/CIPK6 calcium sensor/protein kinase complex. *Cell Research, 21*, 1116–1130.

Jung, J. Y., Shin, R., & Schachtman, D. P. (2009). Ethylene mediates response and tolerance to potassium deprivation in Arabidopsis[W]. *Plant Cell, 21*, 607–621.

Kato, Y., Sakaguchi, M., Mori, Y., Saito, K., Nakamura, T., Bakker, E. P., Sato, Y., Goshima, S., & Uozumi, N. (2001). Evidence in support of a four transmembrane-pore-transmembrane topology model for the Arabidopsis thaliana Na+/K+ translocating AtHKT1 protein, a member of the superfamily of K+ transporters. *Proceedings of the National Academy of Sciences of the United States of America, 98*, 6488–6493.

Kunz, H. H., Gierth, M., Herdean, A., Satoh-Cruz, M., Kramer, D. M., Spetea, C., & Schroeder, J. I. (2014). Plastidial transporters KEA1, −2, and −3 are essential for chloroplast osmoregulation, integrity, and pH regulation in Arabidopsis. *Proceedings of the National Academy of Sciences of the United States of America, 111*, 7480–7485.

Lan, W. Z., Lee, S. C., Che, Y. F., Jiang, Y. Q., & Luan, S. (2011). Mechanistic analysis of AKT1 regulation by the CBL-CIPK-PP2CA interactions. *Molecular Plant, 4*, 527–536.

Lebaudy, A., Very, A. A., & Sentenac, H. (2007). K+ channel activity in plants: Genes, regulations and functions. *FEBS Letters, 581*, 2357–2366.

Lebaudy, A., Hosy, E., Simonneau, T., Sentenac, H., Thibaud, J. B., & Dreyer, I. (2008). Heteromeric K+ channels in plants. *The Plant Journal, 54*, 1076–1082.

Lee, S. C., Lan, W. Z., Kim, B. G., Li, L., Cheong, Y. H., Pandey, G. K., Lu, G., Buchanan, B. B., & Luan, S. (2007). A protein phosphorylation/dephosphorylation network regulates a plant potassium channel. *Proceedings of the National Academy of Sciences of the United States of America, 104*, 15959–15964.

Lejay, L., Wirth, J., Pervent, M., Cross, J. M. F., Tillard, P., & Gojon, A. (2008). Oxidative pentose phosphate pathway-dependent sugar sensing as a mechanism for regulation of root ion transporters by photosynthesis1[W]. *Plant Physiology, 146*, 2036–2053.

Li, L., Kim, B. G., Cheong, Y. H., Pandey, G. K., & Luan, S. (2006). A Ca2+ signaling pathway regulates a K+ channel for low-K response in Arabidopsis. *Proceedings of the National Academy of Sciences of the United States of America, 103*, 12625–12630.

Luan, S., Lan, W., & Chul Lee, S. (2009). Potassium nutrition, sodium toxicity, and calcium signaling: Connections through the CBL-CIPK network. *Current Opinion in Plant Biology, 12*, 339–346.

O'Connell, K. M. S., Rolig, A. S., Whitesell, J. D., & Tamkun, M. M. (2006). Kv2.1 potassium channels are retained within dynamic cell surface microdomains that are defined by a perimeter fence. *The Journal of Neuroscience, 26*, 9609–9618.

Pilot, G., Gaymard, F., Mouline, K., Cherel, I., & Sentenac, H. (2003a). Regulated expression of Arabidopsis shaker K+ channel genes involved in K+ uptake and distribution in the plant. *Plant Molecular Biology, 51*, 773–787.

Pilot, G., Pratelli, R., Gaymard, F., Meyer, Y., & Sentenac, H. (2003b). Five-group distribution of the Shaker-like K+ channel family in higher plants. *Journal of Molecular Evolution, 56*, 418–434.

Poree, F., Wulfetange, K., Naso, A., Carpaneto, A., Roller, A., Natura, G., Bertl, A., Sentenac, H., Thibaud, J. B., & Dreyer, I. (2005). Plant K(in) and K(out) channels: Approaching the trait of opposite rectification by analyzing more than 250 KAT1-SKOR chimeras. *Biochemical and Biophysical Research Communications, 332*, 465–473.

Qi, Z., Hampton, C. R., Shin, R., Barkla, B. J., White, P. J., & Schachtman, D. P. (2008). The high affinity K+ transporter AtHAK5 plays a physiological role in planta at very low K+ concentrations and provides a caesium uptake pathway in Arabidopsis. *Journal of Experimental Botany, 59*, 595–607.

Ragel, P., Rodenas, R., Garcia-Martin, E., Andres, Z., Villalta, I., Nieves-Cordones, M., Rivero, R. M., Martinez, V., Pardo, J. M., Quintero, F. J., & Rubio, F. (2015). The CBL-interacting protein kinase CIPK23 regulates HAK5-mediated high-affinity K+ uptake in Arabidopsis roots. *Plant Physiology, 169*, 2863–2873.

Ragel, P., Raddatz, N., Leidi, E. O., Quintero, F. J., & Pardo, J. M. (2019). Regulation of K+ nutrition in plants. *Frontiers in Plant Science, 10*.

Ren, X. L., Qi, G. N., Feng, H. Q., Zhao, S., Zhao, S. S., Wang, Y., & Wu, W. H. (2013). Calcineurin B-like protein CBL10 directly interacts with AKT1 and modulates K+ homeostasis in Arabidopsis. *The Plant Journal, 74*, 258–266.

Sharma, T., Dreyer, I., & Riedelsberger, J. (2013). The role of K+ channels in uptake and redistribution of potassium in the model plant Arabidopsis thaliana. *Frontiers in Plant Science, 4*.

Shin, R., & Schachtman, D. P. (2004). Hydrogen peroxide mediates plant root cell response to nutrient deprivation. *Proceedings of the National Academy of Sciences of the United States of America, 101*, 8827–8832.

Shin, R., Berg, R. H., & Schachtman, D. P. (2005). Reactive oxygen species and root hairs in Arabidopsis root response to nitrogen, phosphorus and potassium deficiency. *Plant & Cell Physiology, 46*, 1350–1357.

Very, A. A., & Sentenac, H. (2003). Molecular mechanisms and regulation of K+ transport in higher plants. *Annual Review of Plant Biology, 54*, 575–603.

Voelker, C., Schmidt, D., Mueller-Roeber, B., & Czempinski, K. (2006). Members of the Arabidopsis AtTPK/KCO family form homomeric vacuolar channels in planta. *The Plant Journal, 48*, 296–306.

Voelker, C., Gomez-Porras, J. L., Becker, D., Hamamoto, S., Uozumi, N., Gambale, F., Mueller-Roeber, B., Czempinski, K., & Dreyer, I. (2010). Roles of tandem-pore K+ channels in plants – A puzzle still to be solved. *Plant Biology (Stuttgart), 12*(Suppl 1), 56–63.

Wang, Y., Tang, R. J., Yang, X., Zheng, X., Shao, Q., Tang, Q. L., Fu, A., & Luan, S. (2019). Golgi-localized cation/proton exchangers regulate ionic homeostasis and skotomorphogenesis in Arabidopsis. *Plant, Cell & Environment, 42*, 673–687.

Xu, J., Li, H. D., Chen, L. Q., Wang, Y., Liu, L. L., He, L., & Wu, W. H. (2006). A protein kinase, interacting with two calcineurin B-like proteins, regulates K+ transporter AKT1 in Arabidopsis. *Cell, 125*, 1347–1360.

Zhao, S., Zhang, M. L., Ma, T. L., & Wang, Y. (2016). Phosphorylation of ARF2 relieves its repression of transcription of the K+ transporter gene HAK5 in response to low potassium stress[OPEN]. *Plant Cell, 28*, 3005–3019.

Zheng, S., Pan, T., Fan, L., & Qiu, Q. S. (2013). A novel AtKEA gene family, homolog of bacterial K+/H+ antiporters, plays potential roles in K+ homeostasis and osmotic adjustment in Arabidopsis. *PLoS One, 8*, e81463.

Chapter 4
Sequence, Structure and Domain Analysis of Potassium Channels and Transporters

Introduction

Plants have developed a sophisticated and unique mechanism for K⁺ uptake, accumulation and its distribution. Many K⁺ transporters and channels have been identified in the past revealing major breakthroughs of molecular mechanisms involved in their activity and regulation. Most of the K⁺ channels and transporters in plants share structural similarity to animal K⁺ channels and transporters. Considering their same molecular function of K⁺ influx and efflux from prokaryotes to eukaryotes, it is obvious to have a high conservation in their sequence and structural domains. The evolutionary relationship of each channel and transporter family has been deciphered in plants, which provide insights into their conserved regions in different crop species probably sharing similar structures among the plant kingdom. Compared with lower land plants, a dramatic reduction in the diversity of voltage-gated K⁺ channels is seen in the higher plants, which might have been due to transition from aqueous to dry conditions. But these higher plants seem to have developed more channels/transporters that diversified from a single prokaryotic channel/transporter.

Sequence and Domain Architecture

Existence of K⁺ channels in the biological system proposes that they are ancient proteins. At least one K⁺ channel has been identified in almost every bacterial and eukaryotic genome that has been sequenced completely, thus suggesting their ubiquitous presence (Miller, 2000).

© The Author(s), under exclusive license to Springer Nature Switzerland AG 2020
G. K. Pandey, S. Mahiwal, *Role of Potassium in Plants*, SpringerBriefs in Plant
Science, https://doi.org/10.1007/978-3-030-45953-6_4

Characteristics of Potassium Channels and Transporters: Do They Share Common Characteristics?

A striking discovery in the field of K^+ transport has provided greater insights into the atomic basis of preferential selectivity of K^+ channels; till date direct structural information is only limited to bacterial K^+ channel KcsA, which is a K_{ir}-type channel in terms of topology (Kuang et al. 2015). However, electrophysiological experiments suggest that eukaryotic K^+ channels show similar channel characteristics as bacterial channels so one may assume that the structural characteristics of bacterial and eukaryotic channels are also quite similar (Miller 2000). Based on the available crystal structure of bacterial K^+ channels, the information can be extrapolated to eukaryotic K^+ channels. A careful observation suggests that the structure is highly conserved among eukaryotes, as all K^+ channels contain common scaffold carrying the structural motifs determining K^+ selectivity (Hartmann et al. 1991; Yellen et al. 1991; Heginbotham et al. 1994; Doyle et al. 1998). The 'prototype' of K^+ channel majorly consists of structural core and the pore-loop regions comprising of signature sequences responsible for determining the configuration of inner lining of the aqueous pore (Miller 2000). This can be assumed as a 'universal feature' of all K^+ channels (Fig. 4.1). Studies based on hydrophobicity reveal that two transmembrane domains make up the structural core isolated by inwardly pointed pore loops. The structural architecture of the K^+ channels as a whole is such that it looks like a tetrameric structure aligned in fourfold symmetry which may have similar or identical subunits, i.e., they can be homomeric as well as heteromeric (MacKinnon et al. 1993; Miller 2000). The canonical architecture of the K^+ channels may vary based

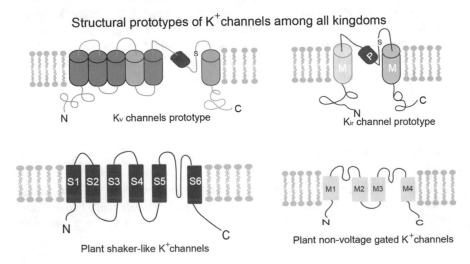

Fig. 4.1 Schematic representation of structural prototype of K channels among all kingdoms. K_v-type and K_{ir}-type channels are present from bacteria to animal. Detailed structure of plant K channels has been also depicted. Plant voltage- and non-voltage-gated channels differ in their structural morphology. (Adapted and modified from Choe 2002)

on their function such as the channels generating response after sensing the change in membrane potential (Choe 2002).

Features of Voltage- and Non-voltage-Gated K+ Channels/ Voltage-Independent K+ Channels in Plant

Voltage-gated K+ channels are characterized by one pore loop and six transmembrane domains (S1–S6), out of which the first four transmembrane domains fold in a particular manner giving rise to voltage sensor module possessing positively charged amino acids (Uozumi et al. 1998; Dreyer and Blatt 2009). The rest of the two TM domains S5 and S6 (fifth and sixth TM domains) along with the pore loop give rise to the foundation of permeation pathway module (Fig. 4.1). Assembly of the four subunits is a major determinant for the formation of inner scaffold of the central transmembrane core and also contributes to the selectivity filter that determines the rectifying properties of the channel (Dreyer and Uozumi 2011).

AtTPK/KCO family shares structural similarity with mammalian 'tandem P-domain channels or K_{ir} channels' having nearly similar secondary structure consisting of four transmembrane domains and two-pore region per subunit (Isayenkov et al. 2011). However, the topology of monomeric subunit of TPK resembles basic motif of a simple K+ channel subunit having 'transmembrane-pore-transmembrane', a tandem repeat structure as a signature (Dreyer et al. 2011). The K+ selectivity of the channel is determined by the 'GYGD signature motif' in the pore region, a key feature of K+ selectivity filters (Fig. 4.1) (Voelker et al. 2006). In the N-terminal region, presence of binding sites of 14-3-3 proteins and variable number of Ca^{2+} binding EF hands in the cytosolic C-terminal region is an indicative of the channel activity being regulated by various mechanisms (Voelker et al. 2010).

Features of Potassium Transporters in Plant

The crystal structure of plant K+ transporters has not been described yet, even after two decades of the identification of the first K+ transporter. However, the structural information based on the homology modeling using amino acid-polyamine-organocation (APC) transporter as template (phylogenetic relationship does exist between APC superfamily and HAK/KT/KUP transporter family) reveals common characteristics of plant transporters (Fig. 4.1) (Vastermark et al. 2014). Characterized by the presence of 10 to 14 transmembrane domains that contributes to the hydrophobic core, the pore region in these transporters has not been traced yet. However, additional cytosolic domains, mainly three in number, are present: N-terminal domain, C-terminal domain and an 70 amino acid stretch in intermediate region between TM2 and TM3 (Ragel et al. 2019). It has been suggested that there are likely chances of interaction between C-terminus and the intermediate loop region of TM2–TM3 anticipating the formation of homodimers (Daras et al. 2015). Studies based on the mutational analysis of these transporters reveal that affinity of the transporter to its substrate can be established by N-terminus and the intermediate region of TM2–TM3 (Santa-Maria et al. 2018).

Evolutionary Relationship of Potassium Channels and Transporters in Plant

In silico analysis tracing the evolutionary history of K⁺ channels/transporters has revealed crucial information about the channels/transporter families in plants. Analysis performed by Gomez-Porras et al. (2012) presented inventory of K⁺ channels/transporter families taking sequences into consideration from moss *Physcomitrella patens*, fern *Selaginella moellendorffii*, monocot *Oryza sativa*, dicots *Arabidopsis thaliana* and the tree *Populus trichocarpa*. Highlights from their study are explained below.

HAK-Type Proteins

In plants, topology of HAK transporters has not been determined experimentally or by in silico predictions, but the hydropathy profile reveals that it consists of 12 putative transmembrane segments with hydrophilic COOH-terminal region. Several consensus motifs have been identified in genome-wide screenings (Gomez-Porras et al., 2012). Phylogenetic analysis using 13 HAKs from *Arabidopsis*, 22 from poplar, 27 from rice, 18 from *P. patens* and 11 from *S. moellendorffii* reveals that 2 HAK5 transporters were the last common ancestor of all embryophytes (Yang et al., 2009). One of these has diverged into a separate group of transporters, whereas another got duplicated at least three times, losing two duplications in lineage ultimately giving rise to tracheophytes forming *P. patens*-specific groups (Gomez-Porras et al. 2012).

HKT-Type Proteins

Plant HKT transporters are members of monovalent cation transporter family which also consists of fungal TRKs and bacterial KtrABs, all sharing common structure of TM-P-TM motifs (Corratge-Faillie et al. 2010). A 30 amino acid-long pore-forming P segment connects two transmembrane helices (Durell and Guy 1999). The common motif structure seems to have evolved from bacterial KscA channel. Consensus motif has been identified in genome-wide screenings of HKT family proteins (Gomez-Porras et al. 2012). Phylogenetic analysis using 1 HKT encoding gene from *Arabidopsis*, 7 from rice, 1 from poplar, 1 from *P. patens* and 6 from *S. moellendorffii* resulted in grouping of all these proteins into a single group of orthologs suggesting that a single HKT protein was the last common ancestor of all embryophytes. Several duplication events in other tracheophytes in different lineages took place in contrast to *P. patens*, which consist of single representative of HKT protein. In rice and *S. moellendorffii*, multiple independent duplications were observed which might have arisen due to their relationship with moist environment (Gomez-Porras et al. 2012).

Shaker-*Type Proteins*

Based on the characteristic pore-forming region, 9 genes coding for *shaker*-like channels in *Arabidopsis*, 11 each in rice and poplar, 4 in *P. patens* and 1 in *S. moellendorffii* were considered for phylogenetic analysis which reveals that a single *shaker*-like K$^+$ channel was present in the ancestor of land plant. Several amplifications occurred in the whole lineage where one amplification is specific to bryophyte and after that a split between the lineages leads to *P. patens* and tracheophytes. Two shaker-like K$^+$ channel genes were present in the tracheophytes, and after the split of angiosperms from this lineage, one of them got lost in the lineage resulting to *S. moellendorffii* (Gomez-Porras et al. 2012).

Voltage-Independent TPK Channels

Phylogenetic analysis taking 6 TPKs from *Arabidopsis*, 3 from rice, 10 from poplar and 3 from *P. patens* together resulted in 2 groups of orthologs implying existence of 2 TPK genes in the common ancestor. Several duplication events took place in two groups, at both species specific and higher levels. One member of this family, i.e., KCO1, slightly deviates from the characteristic feature of the family as it has recently evolved from a very recent evolutionary event of TPK2 gene duplication (Gomez-Porras et al. 2012).

K$^+$/H$^+$ Antiporter Homologs

In *Arabidopsis* genome, six K$^+$ efflux antiporters have been identified. Phylogenetically it is closer to other members of CPA family in *Arabidopsis*. There is no information available for tracing the evolutionary history for these particular antiporters.

Phylogenetic Analysis of K⁺ Transport Machinery

A phylogenetic analysis performed using 98 protein sequences resulted in an unrooted tree as there was no common sequence found among all the proteins used for phylogenetic analysis. The tree was constructed using maximum parsimony method. Neighbour-joining method could not be utilized as there was no common sequence available among them.

Phylogenetic tree constructed by maximum parsimony method (Fig. 4.2) reveals that bacterial KUP, Tok1 (yeast), HsKca1 (human) and HAK5 (*Arabidopsis*) are

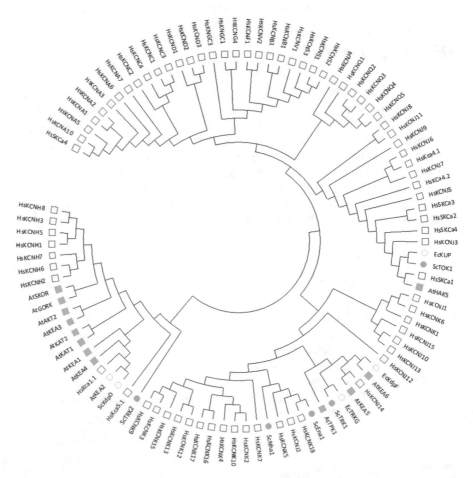

Fig. 4.2 Phylogenetic tree showing association of K⁺ transport system among all the kingdoms of life using sequences from *E. coli*, yeast, human and *Arabidopsis*. Tree constructed by maximum parsimony method results in an unrooted tree as there was no common sequence among the sequences used for this analysis. The tree only depicts phylogenetic relatedness of all the K⁺ transport systems, considering the fact that minimum steps of evolution have taken place while their diversification. Empty rectangular boxes represent proteins from *Homo sapiens*, empty circles represent proteins from *E. coli*, blue circles represent proteins from *S. cerevisiae*, and green rectangular boxes represent sequences from *A. thaliana*

closely related to each other. Other major highlight is the close relationship of bacterial TrkG and TRK1 (yeast) with TPK1 from *Arabidopsis*. All voltage-gated channels from plant and animal form a group probably indicating that they might have evolved parallelly or diversified from a common ancestor.

K+-specific channels and transports share structural similarity consisting of a basic structural prototype among all kingdoms. By looking at the structural motifs and features, one can say that K+ transport systems are conserved structurally as well as functionally from prokaryotes to eukaryotes. It can be hypothesized that they might have evolved from common ancestor. We tried to test this hypothesis by aligning protein sequences from various organisms such as bacteria, yeast, animals and plants, but no common root sequence was found.

References

Choe, S. (2002). Potassium channel structures. *Nature Reviews. Neuroscience, 3*, 115–121.

Corratge-Faillie, C., Jabnoune, M., Zimmermann, S., Very, A. A., Fizames, C., & Sentenac, H. (2010). Potassium and sodium transport in non-animal cells: The Trk/Ktr/HKT transporter family. *Cellular and Molecular Life Sciences, 67*, 2511–2532.

Daras, G., Rigas, S., Tsitsekian, D., Iacovides, T. A., & Hatzopoulos, P. (2015). Potassium transporter TRH1 subunits assemble regulating root-hair elongation autonomously from the cell fate determination pathway. *Plant Science, 231*, 131–137.

Doyle, D. A., Morais Cabral, J., Pfuetzner, R. A., Kuo, A., Gulbis, J. M., Cohen, S. L., Chait, B. T., & MacKinnon, R. (1998). The structure of the potassium channel: Molecular basis of K+ conduction and selectivity. *Science, 280*, 69–77.

Dreyer, I., & Blatt, M. R. (2009). What makes a gate? The ins and outs of Kv-like K+ channels in plants. *Trends in Plant Science, 14*, 383–390.

Dreyer, I., & Uozumi, N. (2011). Potassium channels in plant cells. *The FEBS Journal, 278*, 4293–4303.

Dreyer, I., Plant Biophysics and Heisenberg Group of Biophysics and Molecular Plant Biology, C.d.B.y.G.d.P, Universidad Politécnica de Madrid, Spain, Uozumi, N., Department of Biomolecular Engineering, G.S.o.E, & Tohoku University, Japan. (2011). Potassium channels in plant cells. *The FEBS Journal, 278*, 4293–4303.

Durell, S. R., & Guy, H. R. (1999). Structural models of the KtrB, TrkH, and Trk1,2 symporters based on the structure of the KcsA K(+) channel. *Biophysical Journal, 77*, 789–807.

Gomez-Porras, J. L., Riaño-Pachón, D. M., Benito, B., Haro, R., Sklodowski, K., Rodríguez-Navarro, A., & Dreyer, I. (2012). Phylogenetic analysis of K+ transporters in bryophytes, lycophytes, and flowering plants indicates a specialization of vascular plants. *Frontiers in Plant Science, 3*.

Hartmann, H. A., Kirsch, G. E., Drewe, J. A., Taglialatela, M., Joho, R. H., & Brown, A. M. (1991). Exchange of conduction pathways between two related K+ channels. *Science, 251*, 942–944.

Heginbotham, L., Lu, Z., Abramson, T., & MacKinnon, R. (1994). Mutations in the K+ channel signature sequence. *Biophysical Journal, 66*, 1061–1067.

Isayenkov, S., Isner, J. C., & Maathuis, F. J. (2011). Membrane localization diversity of TPK channels and their physiological role. *Plant Signaling and Behavior, 6*, 1201–1204.

Kuang, Q., Purhonen, P., & Hebert, H. (2015). Structure of potassium channels. *Cellular and Molecular Life Sciences, 72*, 3677–3693.

MacKinnon, R., Aldrich, R. W., & Lee, A. W. (1993). Functional stoichiometry of Shaker potassium channel inactivation. *Science, 262*, 757–759.

Miller, C. (2000). An overview of the potassium channel family. *Genome Biology, 1*, Reviews 0004.

Ragel, P., Raddatz, N., Leidi, E. O., Quintero, F. J., & Pardo, J. M. (2019). Regulation of K+ nutrition in plants. *Frontiers in Plant Science, 10.*

Santa-Maria, G. E., Oliferuk, S., & Moriconi, J. I. (2018). KT-HAK-KUP transporters in major terrestrial photosynthetic organisms: A twenty years tale. *Journal of Plant Physiology, 226,* 77–90.

Uozumi, N., Nakamura, T., Schroeder, J. I., & Muto, S. (1998). Determination of transmembrane topology of an inward-rectifying potassium channel from Arabidopsis thaliana based on functional expression in Escherichia coli. *Proceedings of the National Academy of Sciences of the United States of America, 95,* 9773–9778.

Vastermark, A., Wollwage, S., Houle, M. E., Rio, R., & Saier, M. H., Jr. (2014). Expansion of the APC superfamily of secondary carriers. *Proteins, 82,* 2797–2811.

Voelker, C., Schmidt, D., Mueller-Roeber, B., & Czempinski, K. (2006). Members of the Arabidopsis AtTPK/KCO family form homomeric vacuolar channels in planta. *The Plant Journal, 48,* 296–306.

Voelker, C., Gomez-Porras, J. L., Becker, D., Hamamoto, S., Uozumi, N., Gambale, F., Mueller-Roeber, B., Czempinski, K., & Dreyer, I. (2010). Roles of tandem-pore K+ channels in plants – a puzzle still to be solved. *Plant Biology (Stuttgart), 12*(Suppl 1), 56–63.

Yang, Z., Gao, Q., Sun, C., Li, W., Gu, S., & Xu, C. (2009). Molecular evolution and functional divergence of HAK potassium transporter gene family in rice (Oryza sativa L.). *Journal of Genetics and Genomics, 36,* 161–172.

Yellen, G., Jurman, M. E., Abramson, T., & MacKinnon, R. (1991). Mutations affecting internal TEA blockade identify the probable pore-forming region of a K+ channel. *Science, 251,* 939–942.

Chapter 5
Potassium in Plant Growth and Development

Introduction

Growth and development of an organism is solely dependent on the nutrition. Appropriate amount of nutrients at each developmental stage governs the growth pattern and overall development. K is an important macronutrient for the plant that regulates various physiological processes starting from seed germination to whole plant development which includes nutrient balance, stomatal regulation, photosynthesis, osmoregulation, protein synthesis, cation-anion balance, enzyme activation and stress tolerance.

Potassium in Vegetative Growth and Reproductive Growth

The first stage in the life cycle of the plant is seed germination which starts with the rapid imbibitions of water, and one of the initiators for this process is K (Farooq et al. 2007). One can imagine how important K is to plants as it begins to function at the very early stage of development. K salts such as potassium nitrate (KNO_3), potassium chloride (KCl) and dipotassium hydrogen phosphate (K_2HPO_4) have been proven to be a potent catalyst in improving seed germination as well as emergence rate and often used as seed priming agent. Among all KNO_3 has been shown to be superior osmoprimers exhibiting maximum germination rate in cotton and rice. An optimum K^+/Na^+ ratio is essential to maintain for the proper growth of the plant, as K^+ deficiency or K^+ excess results in hindering overall growth of the plant (Wang et al. 2013; Amanullah et al. 2016). An experiment performed on potato revealed that increasing K levels ranging from 0 to 150 kg ha^{-1} resulted in increase in plant height, aerial stem number and leaf number per plant (Zelelew et al. 2016). Another separate experiment performed on sweet potato suggested reduction in

© The Author(s), under exclusive license to Springer Nature Switzerland AG 2020
G. K. Pandey, S. Mahiwal, *Role of Potassium in Plants*, SpringerBriefs in Plant
Science, https://doi.org/10.1007/978-3-030-45953-6_5

total biomass productivity and root yield when the cultivars were grown on K^+-deficient soil. Almost same effects were observed in cotton: production of plant dry matter, leaf area and size of the internodes ultimately reducing plant growth in vegetative phase of cotton upon K^+ deficiency (Gerardeaux et al. 2008). During the reproductive stage, external application of K^+ plays vital roles in pollen germination and tube growth, and KNO_3 was found to be most effective as an external source of K. A lower external application of K^+ has been shown to delay flowering and physiological maturity. A series of experiments revealed that phenological development (flowering and physiological maturity) is controlled by external K applications in a dose-dependent manner (Amanullah et al. 2016). Split applications of 60 kg ha^{-1} resulted in delayed flowering, whereas a basal dose of 90 kg ha^{-1} resulted in enhancing tasseling, silking and physiological maturity. Split applications of 90 kg ha^{-1} in maize contributed to phenological development in a positive manner (Asif et al. 2007). In conclusion, one can comment that onset of flowering and early physiological maturity depends on externally applied full dose of K^+.

Potassium in Photosynthesis and Grain Filling

Two major factors mainly light and carbon dioxide status are well-known regulators of photosynthesis. However, nutritional status of the plant also determines the photosynthetic efficiency of a plant, as nutrient deficiencies are known to pose a negative impact on it. Several independent studies have demonstrated this in the past. K is known for its role in metabolism and carbohydrate translocation and, thus, also implicated in photosynthesis where it is shown to regulate photosynthesis via sunlight perception (Fig. 5.1) (Hasanuzzaman et al. 2018; Tränkner et al. 2018). When plants are sufficiently supplied with K, sucrose levels in leaves were observed to be several-fold higher, indicating an increase in photosynthesis rate, as higher utilization and export of photoassimilates are known to enhance photosynthesis (Zhao et al. 2001). Amount of K retained in the plant influences the continuous operation of photosynthesis via stomatal regulation (Marschner 2012). K may regulate photosynthesis via different mechanisms. Amount of photosynthates produced is represented by CO_2 entry into intracellular spaces which is regulated by stomatal aperture establishing an indirect role of K in photosynthesis (Hasanuzzaman et al. 2018). Regulation of activity of ATP synthase enzyme by K is another mechanism by which it assists in photosynthesis (Bednarz et al. 1998). Based on observations of K-deficient plants, a direct role of K may be implicated. Under K deficiency conditions, reduction in leaf number and size, reduced stomatal conductance and lower ribulose-1,5-bisphosphate carboxylase/oxygenase (RuBisCO) activity in plants contribute to decreased photosynthesis rate in plants (Zhao et al. 2001). The apparent importance of K in plants was deciphered by molecular identification of K^+ transporters involved in photosynthetic regulation. Three K^+ efflux transporters *AtKEA1/2/3* are shown to regulate primary chloroplast development and

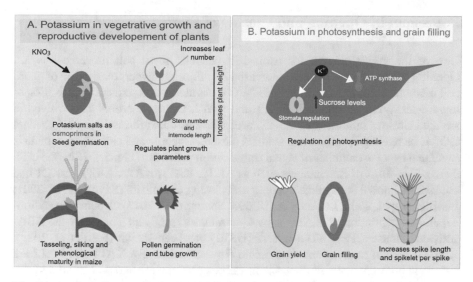

Fig. 5.1 (**a**) K plays an important role in vegetative and reproductive growth of the plants acting as osmoprimer in seed germination as well as pollen germination and also regulates overall plant development at vegetative and reproductive level. (**b**) K increases spike length and spikelet per spike in wheat and also regulates grain filling process. Regulation of photosynthesis occurs via stomatal regulation, ATP synthase regulation and increasing sucrose levels

photosynthesis along with photosynthetic acclimation in fluctuating light environment (Kunz et al. 2014).

Proper applications of K in wheat lead to enhanced photosynthetic rate during grain filling, contributing to a significant increase in grain number and grain filling rate. Higher K content facilitates transportation of resources required for grain filling to developing grains resulting in lower number of sterile grains. Comparatively lower grain sterility, i.e., 22.60%, was observed in *Oryza sativa* upon 100 kg ha^{-1} K application in contrast to higher grain sterility, i.e., 30.33%, without K application (Hasanuzzaman et al. 2018). In wheat, K application under drought stress significantly increased spike length by 21.8%, number of spikelet per spike up to 23.27%, number or grains up to 39.24% and grain yield up to 30.77% (Raza et al. 2014). Another independent report revealed that NPK fertilizer accounts for increasing yield up to 17.6% as compared to NP fertilizer application in wheat, whereas in rice, 1.7–9.8% of increased yield was observed (Duan et al. 2014). Another study observed 13% higher yield in *Triticum aestivum* and 50% higher in *O. sativa* upon 60 kg ha^{-1} of external K application (Khan et al. 2007).

Potassium in Opening and Closing of Stomata

One of the major players for osmosis-dependent guard cell movement is K⁺. Opening of stomata is a result of increased K⁺ influx and hyperpolarization of guard cell membrane through activation of H⁺-ATPases along with sequential activation of inward-rectifying K⁺ channels KAT1 and KAT2. Opposite events are involved in stomata closure; inhibition of H⁺-ATPases activates outward-rectifying K⁺ channel GORK increasing K⁺ efflux and stomata closure. Opening and closing of stomata is regulated by a vast number of signal transduction pathways (Fig. 5.1) (Sharma et al. 2013). Two-third of K⁺ channels such as AKT2, KAT1, KAT2, AKT1, AtKC1 and GORK are known to regulate opening and closing of stomata (Szyroki et al. 2001; Ivashikina et al. 2005; Lebaudy et al. 2008; Sharma et al. 2013). A phosphorylation-dependent regulation of KAT1 at its C-terminus (Thr[306] and Thr[308]) by the ABA-activated kinase OPEN STOMATA1 (OST1) has been shown (Sato et al. 2009). KAT2 is known to contribute to stomata closure exclusively. KAT1 and KAT2 are known to homomerize and dimerize in various permutation and combinations with other members of the shaker channel family for the formation of inward K⁺ channel in guard cell (Lebaudy et al. 2010). Phytohormone abscisic acid (ABA) plays a vital role in this process as it is responsible for the induction of three events in the guard cell, depolarization of guard cell membrane, GORK expression and GORK-mediated K⁺ efflux (Munemasa et al. 2015).

Long-Distance Transport of Potassium

The prerequisite for the long-distance transport of K is uptake by the root system. The first phase of long-distance commences upon movement of K⁺ into the xylem or phloem, whereas successful cellular compartmentalization of K⁺ completes the second phase of long-distance transport (Ahmad and Maathuis 2014). K⁺ absorbed via roots must be loaded into xylem for its transportation to shoots (Park et al. 2008). A key step in long-distance transport is communication between roots to shoot that is mediated by outward-rectifying shaker-like SKOR channel (Gaymard et al. 1998). Expressed mainly in the pericycle and xylem parenchyma, the phyto-hormone ABA can inhibit its expression resulting in obstruction in K⁺ transport to the shoots (Gaymard et al. 1998). The C-terminal transmembrane regions of the SKOR are known for their ability to sense intracellular K⁺ concentration (Liu et al. 2006). However, another report suggested that external K⁺ concentration may also modulate SKOR activity indicating the possibility of a mechanism that enables the SKOR channels to sense K⁺ externally as well as internally. SKOR is under tight regulation of complex interplay between pore region and last TM domain of the channel, which is accountable for opening and closing of the channel (Fig. 5.2) (Sharma et al. 2013). The prerequisite for opening of the channel is highly negative membrane voltage (to prevent unwanted K⁺ influx). Usually in the presence of high

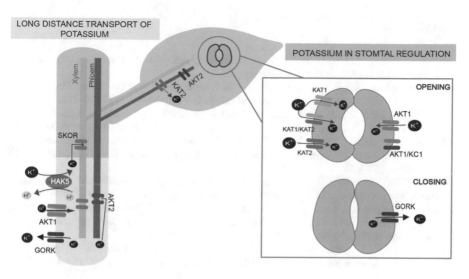

Fig. 5.2 K+ in long-distance transport. GORK, AKT1, HAK5 and SKOR regulate K+ movements in xylem and phloem delivering it to the upper parts of the plants, mainly shoot and leaf. Opening and closing of stomata is also controlled by the K+ fluxes in the guard cell. (Adapted and modified from Adams and Shin 2014)

$[K^+]_{ext,}$ the channel is stabilized in closed state as the rigid pore region interacts with the last transmembrane of the channels. Under low K+ conditions, due to less occupancy of K+ in pore region, it becomes flexible and doesn't allow interaction with adjacent domains (Sharma et al. 2013).

Other transporters also partly play a role in long-distance transport; for example, KAT2, which is a phloem-associated channel, is induced by phytohormone auxin, thus suggesting an important role in K+ homeostasis in phloem (Philippar et al. 2004). Another report suggests that AKT2/3 plays a role in modification of phloem potential and might regulate phloem loading and long-distance transport. Based on previous reports, it was proposed that AKT2/3 and PP2CA might regulate K+ current and membrane potential required for phloem loading and long-distance transport in coordination with each other (Deeken et al. 2002; Adams and Shin 2014). It is pertinent to mention here, a novel role for K+ as an energy source which has been found to operate in plants, where the K+ gradient between two parts serves as a mobile source of energy to take up K+ and sucrose (due to membrane hyperpolarization and change in rectification of channels like AKT2) (Gajdanowicz et al. 2009). This so-called K+ battery operates only for short-term energy limitations and cannot be substituted for high energy requirements.

Essential role of K in various stages of plant development has been discussed briefly in this chapter. Since K+ is absorbed via roots and in order to regulate vital processes in plant, K+ must reach to distal parts of the plants. This is achieved by the long-distance transport of K+ via K+-specific transport machinery.

References

Adams, E., & Shin, R. (2014). Transport, signaling, and homeostasis of potassium and sodium in plants. *Journal of Integrative Plant Biology, 56*, 231–249.

Ahmad, I., & Maathuis, F. J. (2014). Cellular and tissue distribution of potassium: Physiological relevance, mechanisms and regulation. *Journal of Plant Physiology, 171*, 708–714.

Amanullah, I., Irfanullah, A., & Hidayat, Z. (2016). Potassium management for improving growth and grain yield of maize (Zea mays L.) under moisture stress condition. *Sci Rep, 6*, 34627.

Asif, M., Amanullah, & Anwar, M. (2007). Phenology, leaf area and yield of spring maize (Cv. Azam) as affected by levels and timings of potassium application. *World Applied Sciences Journal, 2*, 299–303.

Bednarz, C. W., Oosterhuis, D. M., & Evans, R. D. (1998). Leaf photosynthesis and carbon isotope discrimination of cotton in response to potassium deficiency. *AGRIS, 39*, 131–139.

Deeken, R., Geiger, D., Fromm, J., Koroleva, O., Ache, P., Langenfeld-Heyser, R., Sauer, N., May, S. T., & Hedrich, R. (2002). Loss of the AKT2/3 potassium channel affects sugar loading into the phloem of Arabidopsis I SpringerLink. *Plants, 216*, 334–344.

Duan, Y., Shi, X., Li, S., Sun, X., & He, X. (2014). Nitrogen use efficiency as affected by phosphorus and potassium in long-term rice and wheat experiments. *Journal of Integrative Agriculture, 13*, 588–596.

Farooq, M., Basra, M. A., Rehman, H., & Saleem, B. A. (2007). Seed priming enhances the performance of late sown wheat (Triticum aestivum L.) by improving chilling tolerance. *Journal of Agronomy and Crop Science, 194*, 55–60.

Gajdanowicz, P., Garcia-Mata, C., Gonzalez, W., Morales-Navarro, S. E., Sharma, T., Gonzalez-Nilo, F. D., Gutowicz, J., Mueller-Roeber, B., Blatt, M. R., & Dreyer, I. (2009). Distinct roles of the last transmembrane domain in controlling Arabidopsis K+ channel activity. *The New Phytologist, 182*, 380–391.

Gaymard, F., Pilot, G., Lacombe, B., Bouchez, D., Bruneau, D., Boucherez, J., Michaux-Ferriere, N., Thibaud, J. B., & Sentenac, H. (1998). Identification and disruption of a plant shaker-like outward channel involved in K+ release into the xylem sap. *Cell, 94*, 647–655.

Gerardeaux, E., Meille, L. J., Constantin, J., Pellerin, S., & Dingkuhn, M. (2008). Changes in plant morphology and dry matter partitioning caused by potassium deficiency in Gossypium hirsutum (L.). *Environmental and Experimental Botany, 67*, 451–459.

Hasanuzzaman, M., BBhuyan, M. H. M., Nahar, M. H. M. B., Hossain, K., Mahmud, M. S., Hossen, J. A., Masud, M. S., Moumita, A. A. C., & Fujita, M. (2018). Potassium: A vital regulator of plant responses and tolerance to abiotic stresses. *Agronomy, 8*, 31.

Ivashikina, N., Deeken, R., Fischer, S., Ache, P., & Hedrich, R. (2005). AKT2/3 subunits render guard cell K+ channels Ca2+ sensitive. *The Journal of General Physiology, 125*, 483–492.

Khan, R., Gurmani, A. R., Gurmani, A. K., & Zia, M. S. (2007). Effect of potassium application on CROP yields under wheat- rice system. *Sarhad Journal of Agriculture, 23*.

Kunz, H. H., Gierth, M., Herdean, A., Satoh-Cruz, M., Kramer, D. M., Spetea, C., & Schroeder, J. I. (2014). Plastidial transporters KEA1, −2, and −3 are essential for chloroplast osmoregulation, integrity, and pH regulation in Arabidopsis. *Proceedings of the National Academy of Sciences of the United States of America, 111*, 7480–7485.

Lebaudy, A., Vavasseur, A., Hosy, E., Dreyer, I., Leonhardt, N., Thibaud, J. B., Very, A. A., Simonneau, T., & Sentenac, H. (2008). Plant adaptation to fluctuating environment and biomass production are strongly dependent on guard cell potassium channels. *Proceedings of the National Academy of Sciences of the United States of America, 105*, 5271–5276.

Lebaudy, A., Pascaud, F., Very, A. A., Alcon, C., Dreyer, I., Thibaud, J. B., & Lacombe, B. (2010). Preferential KAT1-KAT2 heteromerization determines inward K+ current properties in Arabidopsis guard cells. *The Journal of Biological Chemistry, 285*, 6265–6274.

Liu, K., Li, L., & Luan, S. (2006). Intracellular K+ sensing of SKOR, a Shaker-type K+ channel from Arabidopsis. *The Plant Journal, 46*, 260–268.

Marschner, P. (2012). *Mineral nutrition of higher plants*. London: Academic.

Munemasa, S., Hauser, F., Park, J., Waadt, R., Brandt, B., & Schroeder, J. I. (2015). Mechanisms of abscisic acid-mediated control of stomatal aperture. *Current Opinion in Plant Biology, 28,* 154–162.

Park, J., Kim, Y. Y., Martinoia, E., & Lee, Y. (2008). Long-distance transporters of inorganic nutrients in plants | SpringerLink. *Journal of Plant Biology, 51,* 240–247.

Philippar, K., Ivashikina, N., Ache, P., Christian, M., Luthen, H., Palme, K., & Hedrich, R. (2004). Auxin activates KAT1 and KAT2, two K+-channel genes expressed in seedlings of Arabidopsis thaliana. *The Plant Journal, 37,* 815–827.

Raza, M. A. S., Saleem, M. F., Shah, G. M., & Khan, I. H. (2014). Exogenous application of glycinebetaine and potassium for improving water relations and grain yield of wheat under drought. *Journal of Soil Science and Plant Nutrition, 14,* 348–364.

Sato, A., Sato, Y., Fukao, Y., Fujiwara, M., Umezawa, T., Shinozaki, K., Hibi, T., Taniguchi, M., Miyake, H., Goto, D. B., & Uozumi, N. (2009). Threonine at position 306 of the KAT1 potassium channel is essential for channel activity and is a target site for ABA-activated SnRK2/OST1/SnRK2.6 protein kinase. *The Biochemical Journal, 424,* 439–448.

Sharma, T., Dreyer, I., & Riedelsberger, J. (2013). The role of K+ channels in uptake and redistribution of potassium in the model plant Arabidopsis thaliana. *Frontiers in Plant Science, 4.*

Szyroki, A., Ivashikina, N., Dietrich, P., Roelfsema, M. R., Ache, P., Reintanz, B., Deeken, R., Godde, M., Felle, H., Steinmeyer, R., Palme, K., & Hedrich, R. (2001). KAT1 is not essential for stomatal opening. *Proceedings of the National Academy of Sciences of the United States of America, 98,* 2917–2921.

Tränkner, M., Tavakol, E., & Jákli, B. (2018). Functioning of potassium and magnesium in photosynthesis, photosynthate translocation and photoprotection. *Physiologia Plantarum, 163,* 414–431.

Wang, M., Zheng, Q., Shen, Q., & Guo, S. (2013). The critical role of potassium in plant stress response. *International Journal of Molecular Sciences, 14,* 7370–7390.

Zelelew, D. Z., Lal, S., Kidane, T. T., & Ghebreslassie, B. M. (2016). Effect of potassium levels on growth and productivity of potato varieties. *American Journal of Plant Sciences, 7,* 1629–1638.

Zhao, M., Oosterhuis, D. M., & Bednarz, C. W. (2001). Influence of potassium deficiency on photosynthesis, chlorophyll content, and chloroplast ultrastructure of cotton plants. *Photosynthetica, 39,* 103–109.

Chapter 6
Potassium in Abiotic Stress

Introduction

A compromise in crop quality and quantity is always seen because of challenged crop health in frequent abiotic stress regimes. Plants have developed a broad range of mechanisms to withstand these stresses. The role of mineral nutrients is greatly established in providing tolerance against abiotic stresses in plants. And K$^+$ availability affects anatomy, morphology and metabolism. Among all nutrients, contribution of K$^+$ is studied exclusively, highlighting the role of K$^+$ in stress tolerance mechanisms.

Potassium in Drought and Waterlogging Stress

A few roles of K in plants provide indication that K and drought might be linked entities. Popularly known for its role in cell turgor maintenance, osmotic adjustment at the cellular level and opening and closing of stomata at the physiological level, a close relationship between drought tolerance and K nutritional status exists. Reports demonstrating this connection successfully suggest that under drought conditions, K sufficiency in the soil leads to improved dry matter of the plant in contrast to low K$^+$ concentration in the soil (Hasanuzzaman et al. 2018). Experiments related to the exogenous application of K reveal that along with its contribution for the improvement of plant dry matter and leaf area, it can also stimulate water uptake upon perception of drought condition (Egilla et al. 2001; Römheld and Kirkby 2010). Although it is a well-established fact that K$^+$ plays a protective role under drought stress, the exact mechanisms explaining this 'K-dependent drought-adaptive response' have not been understood in detail. K-dependent drought-adaptive responses have been reported in several plant species such as *Oryza sativa*, *Triticum*

© The Author(s), under exclusive license to Springer Nature Switzerland AG 2020
G. K. Pandey, S. Mahiwal, *Role of Potassium in Plants*, SpringerBriefs in Plant
Science, https://doi.org/10.1007/978-3-030-45953-6_6

aestivum, Brassica napus, Zea mays and *Phaseolus vulgaris* (Pandey et al. 2004; Din et al. 2011; Jatav et al. 2012; Zhang et al. 2014). These adaptive responses may vary among different crops, starting from maintaining cell turgor, xylem hydraulic conductance and stomatal movements to proline accumulation (plays protective role in osmotic adjustment) (Teixeira and Pereira 2007). Looking at the K-dependent drought-adaptive responses, one may ask if K also shows a protective function in waterlogged conditions. The answer to this question has been already found out by different groups in *Vigna radiata*, wheat and maize. The observations based on the studies performed in these crops can be summarized as follows: the relative water content, turgor potential and leaf water potential were found to be higher in plants supplied with sufficient amount of K exogenously, accompanied by their lower osmotic potential (Gupta and Berkowitz 1987). An obstruction in stomatal conductance, photosynthesis rate and root hydraulic conductivity of plant was prominently seen under waterlogged conditions (Else et al. 2008). Due to the sole requirement of K^+ in opening and closing of stomata, K plays a protective role in both drought stress and waterlogging stress (Humble and Raschke 1971). Though the mechanism behind both these processes is not clear, some reports suggest that stomatal closure in response to waterlogging might reduce photosynthetic efficiency through reduced carboxylation in chloroplast (Pier and Berkowitz 1987). This might justify the large K requirement in water-stressed plants, possibly indicating a role for K in protecting against photo-oxidative damage caused by water stress (Kant and Kafkafi 2002). It was also observed that since large amounts of K are lost from the chloroplast parallel to decrease in photosynthesis, higher K is required to maintain photosynthetic activity of water-stressed plants (Gupta et al. 1989).

Potassium in Low-Temperature Stress

In *Arabidopsis* cold-shock factors are known to be master regulators of chilling or freezing stress; however, the role of nutrients especially K has also been implicated in the chilling stress. Studies suggest that K-regulated mechanisms such as photosynthesis, carbon assimilation and metabolism were found to be downregulated in chilling stress resulting in ROS production and oxidative damage (Hasanuzzaman et al. 2011). Sufficient external K^+ supply can overcome these damages caused by the chilling stress indicating the protective role of K in the process; however, the mechanism still remains unknown (Cakmak 2005). Another report claims that in freezing stress, plants lose water from the apoplast resulting in cold-induced dehydration to which K has been proven to be a 'medicinal herb' as K supply helps in adjusting osmotic potential (Wang et al. 2013). In crop plant maize, one of its cold-sensitive varieties exhibits greater tolerance to chilling stress and better ROS tolerance upon treatment with KCl (Farooq et al. 2007). To summarize, one can say that K is involved in protective mechanisms against chilling stress, as the phenomenon has been observed in several crop species.

Potassium in Salt Stress

The word 'salinity' is mostly linked with sodium (Na^+) in the literature as higher concentrations of Na^+ are present in the saline soils, accompanied by more accumulation of Na^+ in the plants (Tavakkoli et al. 2010). Under salinity stress, whole plant ion levels change, where more Na^+ gets accumulated leading to ion toxicity in the plant cells, which in turn inhibit plant root growth resulting in decrease in nutrient uptake and translocation of K^+ (Shabala and Cuin 2008; Shrivastava and Kumar 2015). In salinity stress, Na^+ competes with K^+ for the binding sites of enzyme complexes in biochemical reactions, ultimately replacing K^+ from the metabolic reactions resulting in disruption of plant metabolism (Botella et al. 1997). NaCl-induced K^+ loss is a common phenomenon under salinity stress, which is more prevalent in salt-sensitive varieties compared to the tolerant varieties (Chen et al. 2007). A hallmark feature of plant salt tolerance is the disruption of cytosolic homeostasis of Na^+/K^+ ratios which is due to the rapid loss of K^+ from the cytosol. This NaCl-induced K^+ efflux from the cytosol to apoplast is mainly mediated by GORK and ROS-activated non-selective cation channels (NSCC) (Jayakannan et al. 2013). Onset of salt stress results in rapid depolarization of plasma membrane activating GORK channels followed by K^+ efflux. Further, more K^+ efflux is induced upon ROS accumulation under stress conditions via activation of ROS-activated NSCC and GORK channels. Retention of K^+ under salt stress conditions in various tissues has been recognized as an important trait for the plant salt tolerance (Dasgan et al. 2002; Hauser and Horie 2010). Studies in various crop plants reveal that K^+ retention ability in roots and leaf tissue seems to be important (Wu et al. 2015). Studies performed on barley and lucerne varieties showed that tolerant varieties have better root K^+ retention ability than their susceptible counterparts. In barley, wheat, poplar, cotton, *Brassica* and *Arabidopsis*, salt-tolerant genotypes exhibit significantly higher K^+ retention ability in mesophyll cells in contrast to the sensitive genotypes (Wu et al. 2013, 2014, 2015; Chakraborty et al. 2016). Hence, cells need to maintain a higher K^+/Na^+ ratio, to reduce the toxic effect of Na^+ and promote plant growth and development. Studies also report that in salinity stress higher application of K^+ reduces Na^+ concentration increasing K^+/Na^+ ratio, thereby increasing K^+ in plant cells (Wang et al. 2013).

Presence of appropriate nutrients in plants aids their tolerance to various abiotic stresses. A complex network of all nutrients might exist to keep a check and balance of overall plant health, ultimately contributing to abiotic stress tolerance. Here in this chapter, we briefly discuss the increasing evidences suggesting broader role of K^+ in plant abiotic stress. Recent advances in K^+ stress biology suggest that K^+ plays protective roles during drought, waterlogging, low temperature and salt stress. Interplay of K^+ and salt stress has been worked out quite well with supporting experimental evidences. Similar studies might shed light on the mechanism of K^+-mediated tolerance during abiotic stress conditions.

References

Botella, M. A., Martinez, V., Pardines, J., & Cerdá, A. (1997). Salinity induced potassium deficiency in maize plants. *Journal of Plant Physiology, 150*, 200–205.

Cakmak, I. (2005). The role of potassium in alleviating detrimental effects of abiotic stresses in plants. *Journal of Plant Nutrition and Soil Science, 168*, 521–530.

Chakraborty, K., Bose, J., Shabala, L., Shabala, S. (2016). Difference in root K retention ability and reduced sensitivity of K-permeable channels to reactive oxygen species confer differential salt tolerance in three species. *Journal of Experimental Botany, 67*(15), 4611–4625.

Chen, Z., Pottosin, I. I., Cuin, T. A., Fuglsang, A. T., Tester, M., Jha, D., Zepeda-Jazo, I., Zhou, M., Palmgren, M. G., Newman, I. A., & Shabala, S. (2007). Root plasma membrane transporters controlling K+/Na+ homeostasis in salt-stressed barley. *Plant Physiology, 145*, 1714–1725.

Dasgan, H. Y., Aktas, H., Abak, K., & Cakmak, I. (2002). Determination of screening techniques to salinity tolerance in tomatoes and investigation of genotype responses. *Plant Science, 163*, 695–703.

Din, J., Khan, S. U., Ali, I., & Gurmani, A. R. (2011). Physiological and agronomic response of canola varieties to drought stress. *AGRIS, 21*.

Egilla, J. N., Davies, F. T., & Drew, M. C. (2001). Effect of potassium on drought resistance of Hibiscus rosa-sinensis cv. Leprechaun: Plant growth, leaf macro- and micronutrient content and root longevity. *Plant and Soil, 229*, 213–224.

Else, M. A., Coupland, D., & Jackson, M. B. (2008). Decreased root hydraulic conductivity reduces leaf water potential, initiates stomatal closure and slows leaf expansion in flooded plants of castor oil (Ricinus communis) despite diminished delivery of ABA from the roots to shoots in xylem sap. *Physiologia Plantarum, 111*, 46–54.

Farooq, M., Basra, M. A., Rehman, H., & Saleem, B. A. (2007). Seed priming enhances the performance of late sown wheat (Triticum aestivum L.) by improving chilling tolerance. *Journal of Agronomy and Crop Science, 194*, 55–60.

Gupta, A. S., & Berkowitz, G. A. (1987). Osmotic adjustment, symplast volume, and nonstomatally mediated water stress inhibition of photosynthesis in wheat. *Plant Physiology, 85*, 1040–1047.

Gupta, A. S., Berkowitz, G. A., & Pier, P. A. (1989). Maintenance of photosynthesis at low leaf water potential in wheat: Role of potassium status and irrigation history. *Plant Physiology, 89*, 1358–1365.

Hasanuzzaman, M., Hossain, M. A., da Silva, J. A. T., & Fujita, M. (2011). Plant response and tolerance to abiotic oxidative stress: Antioxidant Defense is a key factor. In B. Venkateswarlu, C. Shanker, & M. Maheswari (Eds.), *Crop stress and its management: Perspectives and strategies*. Dordrecht: Springer.

Hasanuzzaman, M., BBhuyan, M. H. M., Nahar, M. H. M. B., Hossain, K., Mahmud, M. S., Hossen, J. A., Masud, M. S., Moumita, A. A. C., & Fujita, M. (2018). Potassium: A vital regulator of plant responses and tolerance to abiotic stresses. *Agronomy, 8*, 31.

Hauser, F., & Horie, T. (2010). A conserved primary salt tolerance mechanism mediated by HKT transporters: A mechanism for sodium exclusion and maintenance of high K(+)/Na(+) ratio in leaves during salinity stress. *Plant, Cell & Environment, 33*, 552–565.

Humble, G. D., & Raschke, K. (1971). Stomatal opening quantitatively related to potassium transport: Evidence from electron probe analysis 1. *Plant Physiology, 48*, 447–453.

Jatav, K. S., Agarwal, R. M., Singh, R. P., & Shrivastva, M. (2012). Growth and yield responses of wheat [Triticum aestivum L] to suboptimal water supply and different potassium doses. *Journal of Functional And Environmental Botany, 2*, 39.

Jayakannan, M., Bose, J., Babourina, O., Rengel, Z., & Shabala, S. (2013). Salicylic acid improves salinity tolerance in Arabidopsis by restoring membrane potential and preventing salt-induced K+ loss via a GORK channel. *Journal of Experimental Botany, 64*, 2255–2268.

Kant S, Kafkafi U (2002) Potassium and abiotic stresses in plants (Incomplete)

Pandey, R., Agarwal, R. M., Jeevaratnam, K., & Sharma, G. L. (2004). Osmotic stress-induced alterations in rice (Oryza sativa L.) and recovery on stress release. *Plant Growth Regulation, 42*, 79–87.

Pier, P. A., & Berkowitz, G. A. (1987). Modulation of water stress effects on photosynthesis by altered leaf k. *Plant Physiology, 85*, 655–661.

Römheld, V., & Kirkby, E. A. (2010). Research on potassium in agriculture: Needs and prospects | SpringerLink. *Plant and Soil, 335*, 155–180.

Shabala, S., & Cuin, T. A. (2008). Potassium transport and plant salt tolerance. *Physiologia Plantarum, 133*, 651–669.

Shrivastava, P., & Kumar, R. (2015). Soil salinity: A serious environmental issue and plant growth promoting bacteria as one of the tools for its alleviation. *Saudi Journal of Biological Sciences, 22*, 123–131.

Tavakkoli, E., Rengasamy, P., & McDonald, G. K. (2010). High concentrations of Na+ and Cl– ions in soil solution have simultaneous detrimental effects on growth of faba bean under salinity stress. *Journal of Experimental Botany, 61*, 4449–4459.

Teixeira, J. P., & Pereira, S. (2007). High salinity and drought act on an organ-dependent manner on potato glutamine synthetase expression and accumulation. *AGRIS, 60*, 121–126.

Wang, M., Zheng, Q., Shen, Q., & Guo, S. (2013). The critical role of potassium in plant stress response. *International Journal of Molecular Sciences, 14*, 7370–7390.

Wu, H., Shabala, L., Barry, K., Zhou, M., Shabala, S. (2013) Ability of leaf mesophyll to retain potassium correlates with salinity tolerance in wheat and barley. *Physiologia Plantarum 149*(4), 515–527.

Wu, H., Shabala, L., Zhou, M., & Shabala, S. (2014). Durum and bread wheat differ in their ability to retain potassium in leaf mesophyll: Implications for salinity stress tolerance. *Plant & Cell Physiology, 55*, 1749–1762.

Wu, H., Shabala, L., Zhou, M., & Shabala, S. (2015). Chloroplast-generated ROS dominate NaCl(−) induced K(+) efflux in wheat leaf mesophyll. *Plant Signaling & Behavior, 10*, e1013793.

Zhang, L., Gao, M., Li, S., Alva, A. K., & Ashraf, M. (2014). Potassium fertilization mitigates the adverse effects of drought on selected Zea mays cultivar. *Turkish Journal of Botany, 38*, 713–723.

Chapter 7
Potassium Deficiency: A Stress Signal

Introduction

Naturally, 96% of the soil K transport is driven by most dominant mechanism for K^+ delivery via facilitated diffusion through the soil to the root surface. K availability is also affected by soil density; however, two independent reports have shown contradictory results. One report highlights higher volumetric water content due to soil compaction which results in facilitating transport of K^+ to the root surface. The other report suggests reduction of root length in denser soil emphasizing on the fact that increased K^+ accumulation is not only because of soil density. It is still debatable how exactly plants sense K^+ deficiency. The concept till date is that roots act as sensory organs for K deficiency (Wang and Wu 2010). The other important question is the threshold soil K^+ concentration, below which the plant senses K^+ deficiency. This and many more questions need to be addressed in the near future by experimental proofs to understand K^+ deficiency.

Among all, for agriculturists and breeders, the symptoms of K^+ deficiency and its prevention are of typical interest, and this field has been worked out very well. However, little is known about the mechanism of K^+ deficiency (the details are discussed in another chapter in this book) and the major players regulating it, but more insights are needed to understand the pathways functioning in K deficiency.

Overview of Potassium Deficiency in Plant

K deficiency affects plant in a 'tip-to-toe' manner as K is required in most of the process in each tissue. In K^+ deficiency, several processes that are crucial for the plant are compromised. K deficiency impedes carbon flow as it disturbs the water status of the plant resulting in hampering sink activity (Kanai et al. 2007). K

© The Author(s), under exclusive license to Springer Nature Switzerland AG 2020
G. K. Pandey, S. Mahiwal, *Role of Potassium in Plants*, SpringerBriefs in Plant
Science, https://doi.org/10.1007/978-3-030-45953-6_7

deficiency is known to affect photosynthesis, transpiration and stomatal conductance negatively (Degl'Innocenti et al. 2009; Kanai et al. 2011). In cotton plants, K deficiency results in modification of ultrastructure of chloroplast filled with large starch granules as well as fewer grana with more plastoglobuli giving rise to poor chloroplast ultrastructure and low chlorophyll content associated with reduction in leaf photosynthates (Zhao et al. 2001). The transport of photosynthates from the source to sink tissue is also dramatically reduced under K deficiency (Cakmak 2005; Gerardeaux et al. 2008; Kanai et al. 2011).

Plasma membrane of root epidermal cell is the first to perceive K deficiency signal, which subsequently gets transduced to cytoplasm (Wang and Wu 2010). A series of biochemical and physiological reactions is initiated generating short-term and long-term responses.

Several signaling components such as membrane potential, reactive oxygen species (ROS) and phytohormone are a part of short-term responses that occur within few hours of K^+ deficiency (Wang and Wu 2010). It was observed that even if $[K^+]_{ext}$ is low, no effect on $[K^+]_{cyt}$ was observed during early senescence period. Significant reduction in $[K^+]_{cyt}$ was seen when K^+ deficiency stress continued over days or weeks initiating long-term responses resulting in the initiation of metabolic and morphological changes (Wang and Wu 2013). K deficiency creates a 'hub' of responses in the plants; these responses can be morphological or physiological in nature.

Morphological Responses to K^+ Deficiency

When there is a nutrient stress, there is obvious effect on the root system of the plants due to the change in nutrient availability. Alterations in root architecture have been reported in nutrient stress, mainly low availability of NPK. Low NPK status is well known for its effect on total lateral root length. Few studies have tried to correlate the root system plasticity with K availability. A study based on different genotypes of rice suggests that K-efficient varieties consist of longer roots as compared to susceptible varieties upon subject to low K conditions (Yang et al. 2003). In cotton plants, increased root length and root surface area were positively correlated to K accumulation (Xiao-Li et al. 2008). Also upon comparing different cotton varieties, it was suggested that more root surface area was observed in K^+-efficient cultivar as compared to K^+-sensitive cultivars (Fig. 7.1) (Brouder and Kenneth 1990; Hua et al. 2009). In model plant *Arabidopsis thaliana* and barley, K deficiency is known to generate a complex set of processes including increase in ROS and ethylene levels and differential synthesis of auxin, cytokinins, etc. that might in turn lead to change in the root system architecture (RSA) (Jung et al. 2009; Nam et al. 2012; Li et al. 2017). The most prominent change in RSA is displayed through the decrease in number of lateral roots and lateral root development accompanied by increase in root hair length (Fig. 7.1) (Jung et al. 2009; Drew 1975; Shin et al. 2005; Amtmann et al. 2008). Studies based on natural variation of root traits (primary root length,

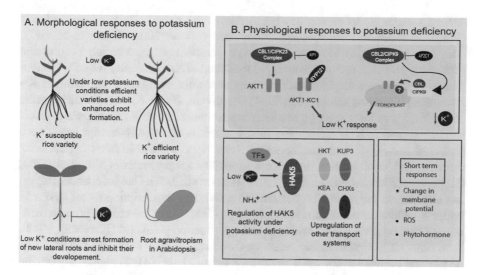

Fig. 7.1 (**a**) Morphological changes induced by K deficiency; root agravitropism, lateral root inhibition and enhanced root formation are the major phenotypic responses of K deficiency. (**b**) Physiological responses can be short term such as change in membrane potential, ROS generation and phytohormone responses. Regulation of channels and transporters are a part of quick as well as delayed responses

lateral root length and root size) among 26 *Arabidopsis* accessions reveal different strategies employed by these accessions in limiting K^+ condition (Kellermeier et al. 2013). In *Arabidopsis* roots, gravitropic behaviour is known to be affected in low K media, through the action of K^+ carrier tiny root hair 1 (TRH1), which also transports auxin required for gravitropic growth of plants (Vicente-Agullo et al. 2004).

Physiological Responses to Potassium Deficiency

Root K^+ absorption depends upon the co-occurrence as well as activity of K^+ transport systems and the proton pumps (which are generally H^+-ATPases) on the plasma membrane (Minjian et al. 2007). It is well demonstrated in tomato that low pH around the surface of the root coupled to net high K^+ influx is associated with overall K^+ uptake efficiency (Chen and Gabelman 2000). Plasma membrane (PM) H^+-ATPases are implicated in K^+ unloading from xylem stream, as specific isoforms of PM H^+-ATPases are upregulated in vessel-associated cells of wood parenchyma, in young stems of poplar plants (Arend et al. 2004).

At the physiological level, one of the major responses to K^+ deficiency is the regulation of K^+ transport proteins since K starvation is known to induce K^+ uptake in plant cell (Shin and Schachtman 2004). In order to ensure K^+ uptake in K deficiency, expression of high-affinity transporters is readily induced that also is thought to serve as a major adaptation mechanism in K^+ starvation. High inducibility of

AtHAK5 under K deficiency has been reported by several studies independently (Fig. 7.1) (Shin and Schachtman 2004; Ashley et al. 2005; Gierth et al. 2005; Amtmann et al. 2008). However, separate experiments suggest variable expression profile for *AtHAK5*, ranging from rapid induction of *AtHAK5* after 6 h of K^+ starvation to gradual increase in transcript abundance after 48 h supplementation of K^+-free media (Shin and Schachtman 2004; Gierth et al. 2005). Also, expression profile of *AtHAK5* was influenced by media compositions used. *AtHAK5* transcript levels were repressed in NH_4^+-containing media, whereas elevated transcript levels were observed in NH_4^+-free media. These results indicate a possible role of NH_4^+ in K^+ nutrition and also question the selectivity of K^+ sensing system (Ahn et al. 2004; Gierth et al. 2005; Amtmann et al. 2008; Rubio et al. 2008). One independent report has also shown the regulation of *AtHAK5* activity by transcription factor RAP2.11 under K deficiency conditions indicating a possible role of the transcription in regulating other components of low K signal transduction pathways (Kim et al. 2012). Four other transcription factors, namely, dwarf and delayed flowering (DDF2), jagged lateral organs (JLO), basic helix-loop-helix121 (bHLH121) and transcription factor II_A gamma chain (TFII_A), have been identified to be upregulated by K deficiency (Hong et al. 2013). Several individual reports have highlighted that abundance of *HAK5* orthologs is greatly enhanced upon K deprivation in barley, tomato and *Oryza sativa* (*OsHAK1*) and *Capsicum annum* (*CaHAK1*) (Banuelos et al. 2002; Rubio et al. 2008). In *Nicotiana rustica*, *NrHAK1* was shown to be upregulated in K starvation specifically expressing at root tip (Guo et al. 2008). In barley, HKT transporters are also shown to be upregulated at transcriptional level under K deficiency (Wang et al. 1998). After observing this trend in different crop species, activation of HAK5 in response to K deficiency is thought as a common feature shared among various plant families (Ashley et al. 2005). Other K transporter such as *AtKUP3* is also shown to be induced under K deficiency along with surprising repression of *AtKUP2* expression (Kim et al. 1998). Transporters from other families such as KEA transporter family and CHX family are also known to be induced by K deprivation after 6 h in *A. thaliana* (Fig. 7.1) (Cellier et al. 2004; Shin and Schachtman 2004).

Under K deficiency, K^+ channels are regulated transcriptionally as well as post-translationally. *AtAKT1* is known to be regulated post-translationally by CIPK23 via phosphorylation mechanism. Using electrophysiological techniques, AKT1 activity is also known to be regulated by AtKC1, as formation of its heterotetrameric complex with AKT1 negatively shifts the voltage dependence of the channel resulting in inhibition of AKT1-mediated K^+ currents, limiting K^+ transport through AKT1 under K deficiency (Duby et al. 2008; Geiger et al. 2009; Wang and Wu 2010). Another study gave an important insight into this heteromeric AKT1-KC1 complex; selective interaction of AtKC1 with SYP121 has been shown to form tripartite K^+ channel complex, which has an important role in K deficiency (Honsbein et al. 2009). AtSKOR and AtAKT2 have been known to be downregulated at transcription level upon K starvation, due to their involvement in long-distance transport. It is speculated that in order to restrict the movement of K^+ between tissues and organs, their reduced expression is utilized to commute the K status from root to shoot (Maathuis et al. 2003; Pilot et al. 2003).

The vacuole is an important K^+ pool regulating cytoplasmic K^+ homeostasis via K^+ accumulation or release into cytoplasm (Walker et al. 1996). Another low K signaling pathway has been proposed in *Arabidopsis* implicating the involvement of tonoplast-localized calcineurin B-like protein (CBL) interacting kinase 9 (CIPK9) (Pandey et al. 2007). The root growth of the *cipk9* mutant plants showed hypersensitive growth behaviour in K^+-deficient media (Pandey et al. 2007). A tonoplast-localized calcineurin B-like protein (CBL2 or CBL3) forms a complex with CIPK9 and localizes to tonoplast (Liu et al. 2013). Based on this it can be assumed that this complex might be regulating an unknown K^+ transport protein at the vacuolar membrane in low K^+ condition. Recent reports suggest that AP2C1 (a type 2C phosphatase) negatively regulates K^+ deficiency response via dephosphorylating CIPK9 (Singh et al. 2018). This study highlights that the possible presence of an alternative module consisting of CIPK9-AP2C1 might be working in parallel to CIPK23-AKT1-AIP1 module (Srivastava et al. 2019). Till date, the targets of CIPK9 have not been identified, and as speculated, it might form an alternative pathway (than the already known CBL1/9-CIPK23-AKT1) regulating K^+ homeostasis (Pandey 2008; Pandey et al. 2014). This phenomenon is important to regulate K^+ homeostasis between these two compartments under K^+-deficient conditions.

Symptoms of Potassium Deficiency

One interesting fact about K deficiency is that the symptoms appear on the older leaves first as compared to younger leaves (Fageria et al. 2002). Plants appear dark green with yellowish brown or dark brown in K deficiency. Necrotic spots first appear on the tips of older leaves, resulting in yellowish-brown leaf tips. Upon prolonged deficiency, older leaves change from yellow to brown, and subsequently, discoloration gradually appears on younger leaves as well (Tiwari 2002). This is thought to be a survival mechanism of the plant in K-deficient condition where remobilization of K^+ from older and senescing organ to younger organ takes place (Cochrane and Cochrane 2009). In a single leaf, leaf edge is affected at first, with the deficiency symptoms gradually appearing on the leaf base. As studied so far, veinal chlorosis is the mainly observed symptom for K deficiency which can be related to chlorophyll degradation (Fageria et al. 2002). Other symptoms include curling of leaf tips and brown scorching.

Potassium Deficiency: A Stress Signal?

K^+ fluxes in the plants lead to alter functioning of metabolic processes, ROS homeostasis and post-translational modifications such as phosphorylation/dephosphorylation (Shin and Schachtman 2004; Ma et al. 2012). K -dependent signaling cascades have been identified in bacteria where K-limiting conditions tend to switch on high-affinity transport system. However, no similar signaling cascade has been identified in plants, but there are speculations that K deficiency in plants may act as a stress signal, switching on various defence responses, to withstand K deficiency (Ashley et al. 2005). First line of indication was provided by transcriptomics study, which reveals that genes related to jasmonic acid signaling and metabolism are largely affected upon K^+ deprivation (Armengaud et al. 2004). Based on this, it can be assumed that low K condition somehow triggers the expression of these genes. The second demonstration showing involvement of root hair defective 2 (*RHD2*), which is an NADPH oxidase generating H_2O_2 in response to K deprivation further suggests activation of ROS-dependent signaling pathway (Foreman et al. 2003). The third line of indication came from direct measurement of ethylene released by K-deficient and K-sufficient plants. More ethylene production by K-starved plants point to triggering of ethylene-mediated signaling in response to K starvation (Shin and Schachtman 2004). Another report demonstrated that Trp-dependent auxin biosynthetic genes (CYP79B2 and CYP79B3) were downregulated in K-starved roots upon K supplementation. This further suggested the role of auxin-dependent process in the fine-tuning of K deficiency responses (Armengaud et al. 2004).

Involvement of Calcium Signaling in K^+ Deprivation

Ca^{2+} signaling, which can be stated as a pivotal signaling system in plant cell, gets activated in response to various environmental stresses as well as nutrient deprivation (Pandey 2008; Wang and Wu 2010). Earlier reports demonstrated that H_2O_2 production under low K conditions is associated with changes in intracellular Ca^{2+} fluxes (Foreman et al. 2003; Shin and Schachtman 2004). ROS and Ca^{2+} sensors have been shown to play an important role in low K sensing and signaling (Xu et al. 2006). NADPH oxidase-mediated ROS production can be stimulated upon elevation in Ca^{2+} concentration in cytoplasm (Yang and Poovaiah 2002). This leads to further Ca^{2+} influx via activation of Ca^{2+}-permeable ion channels (Demidchik et al. 2007). Therefore, it was suggested that Ca^{2+} may act as a second messenger in response to K deficiency (Hafsi et al. 2014). Plasma membrane-localized NADPH oxidases are targeted by either OST1 or CBL1/CBL9-CIPK26 resulting in ROS production (Drerup et al. 2013).

Recent reports have suggested that two successively distinct Ca^{2+} signals in roots are generated upon onset of K^+ deficiency. These two signals displaying spatial and temporal specificity convey responses to K^+ deficiency (Behera et al. 2017). Low K^+ condition induces hyperpolarization of plasma membrane resulting in activation of Ca^{2+} channels present on root epidermis and root hair zone (Very and Davies 2000). In addition to this, other Ca^{2+} channels present within inner membranes can be activated by increased Ca^{2+} in the cytosol, thereby causing Ca^{2+} release from the Ca^{2+} pools such as vacuole and ER (Wang et al. 2018). Further Ca^{2+} influx is mediated by

Ca^{2+}-permeable ion channels which get activated by NADPH oxidase-dependent ROS production (Pei et al. 2000; Demidchik et al. 2007). ROS along with Ca^{2+} is also thought to be involved in long-distance signaling, probably for generating Ca^{2+} waves (Choi et al. 2017). However, it is still not clear how transient increase in intracellular Ca^{2+}, i.e., $[Ca^{2+}]_i$, is sensed by plant cell in response to low K+ stress. As far as the perception of Ca^{2+} signals is concerned, the obvious choice for decoding Ca^{2+} signal in a cell is Ca^{2+} sensor, such as calmodulin (CaM), calmodulin-like proteins (CML), calcineurin B-like proteins (CBL) and Ca^{2+}-dependent protein kinases (CDPK). These Ca^{2+} sensors differ in their mode of sensing Ca^{2+} as well as relay mechanisms. Sensor responders such as CDPK function via intramolecular interactions where Ca^{2+}-induced conformational changes alter their structure and activity. On the other hand, sensor relays such as CaM, CBL and CML function via bimolecular interactions where, at first, sensors combine with Ca^{2+} and, subsequently, relay signal to an interacting partner via conformational change (Sanders et al. 2002; Hashimoto and Kudla 2011).

Few reports highlight the role of CDPKs and show that CDPKs are also known to be involved in regulating K+ channels. CPK11 and CPK24 together mediate Ca^{2+}-dependent inhibition of the activity of shaker pollen inward K+ channels (SPIK/AKT6) in pollen tubes (Zhao et al. 2013). CPK10 has been shown to be involved in the Ca^{2+}-dependent inhibition of K^+_{in} channels in guard cells (Zou et al. 2010). Specific inhibition of guard cell-expressed KAT2 and KAT1 shaker K+ channels is brought about by CPK13 (Ronzier et al. 2014).

Another sensor relay protein such as CaM-related proteins (CML) is involved in coordinating plant responses to abiotic and biotic stress in addition to their role in signaling 'cross talk' (Bender et al. 2013; Ranty et al. 2016; Wu et al. 2017). Members of CML family have been reported to play a vital role in Ca^{2+}-mediated plant responses to K+ deficiency. CML25, a member of CML family, has been shown to be an important transducer that is involved in the Ca^{2+}-mediated regulation of K+ influx (Wang et al. 2015). In addition, a Raf-like MAPKK kinase (AtILK1) directly interacts with AtHAK5 in conjunction with the AtCML9, promoting AtHAK5 accumulation on the membrane (Brauer et al. 2016). These results indicate that Ca^{2+} sensor proteins might play vital roles in connecting Ca^{2+} signaling and downstream target proteins during plant responses to K+ deficiency (Bender et al. 2013).

Another protein module consisting of CBLs and CIPKs also regulates plant responses to K+ deficiency by modulating the activity of AKT1. This mechanism has been discussed briefly earlier in Chap. 3.

Current research area of K+ biology is focussed upon studying K+ deficiency at the cellular and molecular level. Studies have come up with extensive information about the K+ deficiency responses at morphological as well as physiological levels in plants. A new spotlight on K+ deprivation has shown involvement of Ca^{2+} signaling and ROS, implicating the role of these signals in low K+ sensing and signaling.

References

Ahn, S. J., Shin, R., & Schachtman, D. P. (2004). Expression of KT/KUP genes in Arabidopsis and the role of root hairs in K+ uptake. *Plant Physiology, 134*, 1135–1145.

Amtmann, A., Troufflard, S., & Armengaud, P. (2008). The effect of potassium nutrition on pest and disease resistance in plants. *Physiologia Plantarum, 133*, 682–691.

Arend, M., Monshausen, G., Wind, C., Weisenseel, M. H., & Fromm, J. (2004). Effect of potassium deficiency on the plasma membrane H+-ATPase of the wood ray parenchyma in poplar. *Plant, Cell & Environment, 27*, 1288–1296.

Armengaud, P., Breitling, R., & Amtmann, A. (2004). The potassium-dependent transcriptome of Arabidopsis reveals a prominent role of jasmonic acid in nutrient signaling. *Plant Physiology, 136*, 2556–2576.

Ashley, M. K., Grant, M., & Grabov, A. (2005). Plant responses to potassium deficiencies: a role for potassium transport proteins. *Journal of Experimental Botany, 57*, 425–436.

Banuelos, M. A., Garciadeblas, B., Cubero, B., & Rodriguez-Navarro, A. (2002). Inventory and functional characterization of the HAK potassium transporters of rice. *Plant Physiology, 130*, 784–795.

Behera, S., Long, Y., Schmitz-Thom, I., Wang, X. P., Zhang, C., Li, H., Steinhorst, L., Manishankar, P., Ren, X. L., Offenborn, J. N., Wu, W. H., Kudla, J., & Wang, Y. (2017). Two spatially and temporally distinct Ca(2+) signals convey Arabidopsis thaliana responses to K(+) deficiency. *The New Phytologist, 213*, 739–750.

Bender, K. W., Rosenbaum, D. M., Vanderbeld, B., Ubaid, M., & Snedden, W. A. (2013). The Arabidopsis calmodulin-like protein, CML39, functions during early seedling establishment. *The Plant Journal, 76*, 634–647.

Brauer, E. K., Ahsan, N., Dale, R., Kato, N., Coluccio, A. E., Pineros, M. A., Kochian, L. V., Thelen, J. J., & Popescu, S. C. (2016). The Raf-like kinase ILK1 and the high affinity K+ transporter HAK5 are required for innate immunity and abiotic stress response. *Plant Physiology, 171*, 1470–1484.

Brouder, S. M., & Kenneth, C. (1990). Root development of two cotton cultivars in relation to potassium uptake and plant growth in a vermiculite soil. *Field Crops Research, 23*, 187–203.

Cakmak, I. (2005). The role of potassium in alleviating detrimental effects of abiotic stresses in plants. *Journal of Plant Nutrition and Soil Science, 168*, 521–530.

Cellier, F., Conejero, G., Ricaud, L., Luu, D. T., Lepetit, M., Gosti, F., & Casse, F. (2004). Characterization of AtCHX17, a member of the cation/H+ exchangers, CHX family, from Arabidopsis thaliana suggests a role in K+ homeostasis. *The Plant Journal, 39*, 834–846.

Chen, J. J., & Gabelman, W. H. (2000). Morphological and physiological characteristics of tomato roots associated with potassium-acquisition efficiency. *Scientia Horticulturae, 83*, 213–225.

Choi, W. G., Miller, G., Wallace, I., Harper, J., Mittler, R., & Gilroy, S. (2017). Orchestrating rapid long-distance signaling in plants with Ca(2+), ROS and electrical signals. *The Plant Journal, 90*, 698–707.

Cochrane, T. T., & Cochrane, T. A. (2009). The vital role of potassium in the osmotic mechanism of stomata aperture modulation and its link with potassium deficiency. *Plant Signaling and Behavior, 4*, 240–243.

Degl'Innocenti, E., Hafsi, C., Guidi, L., & Navari-Izzo, F. (2009). The effect of salinity on photosynthetic activity in potassium-deficient barley species. *Journal of Plant Physiology, 166*, 1968–1981.

Demidchik, V., Shabala, S. N., & Davies, J. M. (2007). Spatial variation in H2O2 response of Arabidopsis thaliana root epidermal Ca2+ flux and plasma membrane Ca2+ channels. *The Plant Journal, 49*, 377–386.

Drerup, M. M., Schlucking, K., Hashimoto, K., Manishankar, P., Steinhorst, L., Kuchitsu, K., & Kudla, J. (2013). The Calcineurin B-like calcium sensors CBL1 and CBL9 together with their interacting protein kinase CIPK26 regulate the Arabidopsis NADPH oxidase RBOHF. *Molecular Plant, 6*, 559–569.

Drew, M. C. (1975). Comparison of the effects of a localised supply of phosphate, nitrate, ammonium and potassium on the growth of the seminal root system, and the shoot, in barley. *New Phytologist, 75*, 479–490.

Duby, G., Hosy, E., Fizames, C., Alcon, C., Costa, A., Sentenac, H., & Thibaud, J. B. (2008). AtKC1, a conditionally targeted Shaker-type subunit, regulates the activity of plant K+ channels. *The Plant Journal, 53*, 115–123.

Fageria, N. K., Barbosa-Filho, M. P., & Da-Costa, J. G. C. (2002). Potassium use efficiency in common bean genotypes. *Journal of Plant Nutrition, 24*, 1937–1945.

Foreman, J., Demidchik, V., Bothwell, J. H., Mylona, P., Miedema, H., Torres, M. A., Linstead, P., Costa, S., Brownlee, C., Jones, J. D., Davies, J. M., & Dolan, L. (2003). Reactive oxygen species produced by NADPH oxidase regulate plant cell growth. *Nature, 422*, 442–446.

Geiger, D., Becker, D., Vosloh, D., Gambale, F., Palme, K., Rehers, M., Anschuetz, U., Dreyer, I., Kudla, J., & Hedrich, R. (2009). Heteromeric AtKC1·AKT1 channels in Arabidopsis roots facilitate growth under K+-limiting conditions*. *The Journal of Biological Chemistry, 284*, 21288–21295.

Gerardeaux, E., Meille, L. J., Constantin, J., Pellerin, S., & Dingkuhn, M. (2008). Changes in plant morphology and dry matter partitioning caused by potassium deficiency in Gossypium hirsutum (L.). *Environmental and Experimental Botany, 67*, 451–459.

Gierth, M., Mäser, P., & Schroeder, J. I. (2005). The potassium transporter AtHAK5 functions in K+ deprivation-induced high-affinity K+ uptake and AKT1 K+ channel contribution to K+ uptake kinetics in Arabidopsis roots1[w]. *Plant Physiology, 137*, 1105–1114.

Guo, Z. K., Yang, Q., Wan, X. Q., & Yan, P. Q. (2008). Functional characterization of a potassium transporter gene NrHAK1 in Nicotiana rustica. *Journal of Zhejiang University. Science. B, 9*, 944–952.

Hafsi, C., Debez, A., & Abdelly, C. (2014). Potassium deficiency in plants: Effects and signaling cascades | SpringerLink. *Acta Physiologiae Plantarum, 36*, 1055–1070.

Hashimoto, K., & Kudla, J. (2011). Calcium decoding mechanisms in plants. *Biochimie, 93*, 2054–2059.

Hong, J. P., Takeshi, Y., Kondou, Y., Schachtman, D. P., Matsui, M., & Shin, R. (2013). Identification and characterization of transcription factors regulating Arabidopsis HAK5. *Plant and Cell Physiology, 54*, 1478–1490.

Honsbein, A., Sokolovski, S., Grefen, C., Campanoni, P., Pratelli, R., Paneque, M., Chen, Z., Johansson, I., & Blatt, M. R. (2009). A tripartite SNARE-K+ channel complex mediates in channel-dependent K+ nutrition in Arabidopsis. *Plant Cell, 21*, 2859–2877.

Hua, H.-B., Li, Z.-H., & Tian, X. L. (2009). Mechanism of tolerance to potassium deficiency between Liaomian 18 and NuCOTN99B at seedling stage. *Acta Agronomica Sinica, 35*, 475–482.

Jung, J. Y., Shin, R., & Schachtman, D. P. (2009). Ethylene mediates response and tolerance to potassium deprivation in Arabidopsis[W]. *Plant Cell, 21*, 607–621.

Kanai, S., Ohkura, K., Adu-Gyamfi, J. J., Mohapatra, P. K., Nguyen, N. T., Saneoka, H., & Fujita, K. (2007). Depression of sink activity precedes the inhibition of biomass production in tomato plants subjected to potassium deficiency stress. *Journal of Experimental Botany, 58*, 2917–2928.

Kanai, S., Moghaieb, R. E., El-Shemy, H. A., Panigrahi, R., Mohapatra, P. K., Ito, J., Nguyen, N. T., Saneoka, H., & Fujita, K. (2011). Potassium deficiency affects water status and photosynthetic rate of the vegetative sink in green house tomato prior to its effects on source activity. *Plant Science, 180*, 368–374.

Kellermeier, F., Chardon, F., & Amtmann, A. (2013). Natural variation of Arabidopsis root architecture reveals complementing adaptive strategies to potassium starvation1[C][W][OA]. *Plant Physiology, 161*, 1421–1432.

Kim, E. J., Kwak, J. M., Uozumi, N., & Schroeder, J. I. (1998). AtKUP1: An Arabidopsis gene encoding high-affinity potassium transport activity. *Plant Cell, 10*, 51–62.

Kim, M. J., Ruzicka, D., Shin, R., & Schachtman, D. P. (2012). The Arabidopsis AP2/ERF transcription factor RAP2.11 modulates plant response to low-potassium conditions. *Molecular Plant, 5*, 1042–1057.

Li, J., Wu, W. H., & Wang, Y. (2017). Potassium channel AKT1 is involved in the auxin-mediated root growth inhibition in Arabidopsis response to low K(+) stress. *Journal of Integrative Plant Biology, 59*, 895–909.

Liu, L. L., Ren, H. M., Chen, L. Q., Wang, Y., & Wu, W. H. (2013). A protein kinase, calcineurin B-like protein-interacting protein Kinase9, interacts with calcium sensor calcineurin B-like Protein3 and regulates potassium homeostasis under low-potassium stress in Arabidopsis. *Plant Physiology, 161*, 266–277.

Ma, T. L., Wu, W. H., & Wang, Y. (2012). Transcriptome analysis of rice root responses to potassium deficiency. *BMC Plant Biology, 12*, 161.

Maathuis, F. J., Filatov, V., Herzyk, P., Krijger, G. C., Axelsen, K. B., Chen, S., Green, B. J., Li, Y., Madagan, K. L., Sanchez-Fernandez, R., Forde, B. G., Palmgren, M. G., Rea, P. A., Williams, L. E., Sanders, D., & Amtmann, A. (2003). Transcriptome analysis of root transporters reveals participation of multiple gene families in the response to cation stress. *The Plant Journal, 35*, 675–692.

Minjian, C., Haiqiu, Y., Hongkui, Y., & Chunji, J. (2007). Difference in tolerance to potassium deficiency between two maize inbred lines. *Plant Production Science, 10*, 42–46.

Nam, Y. J., Tran, L. S., Kojima, M., Sakakibara, H., Nishiyama, R., & Shin, R. (2012). Regulatory roles of cytokinins and cytokinin signaling in response to potassium deficiency in Arabidopsis. *PLoS One, 7*, e47797.

Pandey, G. K. (2008). Emergence of a novel calcium signaling pathway in plants: CBL-CIPK signaling network. *Physiology and Molecular Biology of Plants, 14*, 51–68.

Pandey, G. K., Cheong, Y. H., Kim, B. G., Grant, J. J., Li, L., & Luan, S. (2007). CIPK9: A calcium sensor-interacting protein kinase required for low-potassium tolerance in Arabidopsis. *Cell Research, 17*, 411–421.

Pandey GK, Kanwar P, Pandey (2014). Global comparative analysis of CBL-CIPK gene families in plants. *SpringerBriefs in Plant Science* ISBN-13: 978-3319090771.

Pei, Z. M., Murata, Y., Benning, G., Thomine, S., Klusener, B., Allen, G. J., Grill, E., & Schroeder, J. I. (2000). Calcium channels activated by hydrogen peroxide mediate abscisic acid signalling in guard cells. *Nature, 406*, 731–734.

Pilot, G., Gaymard, F., Mouline, K., Cherel, I., & Sentenac, H. (2003). Regulated expression of Arabidopsis shaker K+ channel genes involved in K+ uptake and distribution in the plant. *Plant Molecular Biology, 51*, 773–787.

Ranty, B., Aldon, D., Cotelle, V., Galaud, J. P., Thuleau, P., & Mazars, C. (2016). Calcium sensors as key hubs in plant responses to biotic and abiotic stresses. *Frontiers in Plant Science, 7*, 327.

Ronzier, E., Corratge-Faillie, C., Sanchez, F., Prado, K., Briere, C., Leonhardt, N., Thibaud, J. B., & Xiong, T. C. (2014). CPK13, a noncanonical Ca^{2+}–dependent protein kinase, specifically inhibits KAT2 and KAT1 shaker K+ channels and reduces stomatal opening. *Plant Physiology, 166*, 314–326.

Rubio, F., Nieves-Cordones, M., Aleman, F., & Martinez, V. (2008). Relative contribution of AtHAK5 and AtAKT1 to K+ uptake in the high-affinity range of concentrations. *Physiologia Plantarum, 134*, 598–608.

Sanders, D., Pelloux, J., Brownlee, C., & Harper, J. F. (2002). Calcium at the crossroads of signaling. *Plant Cell, 14*(Suppl), S401–S417.

Shin, R., & Schachtman, D. P. (2004). Hydrogen peroxide mediates plant root cell response to nutrient deprivation. *Proceedings of the National Academy of Sciences of the United States of America, 101*, 8827–8832.

Shin, R., Berg, R. H., & Schachtman, D. P. (2005). Reactive oxygen species and root hairs in Arabidopsis root response to nitrogen, phosphorus and potassium deficiency. *Plant & Cell Physiology, 46*, 1350–1357.

Singh, A., Yadav, A. K., Kaur, K., Sanyal, S. K., Jha, S. K., Fernandes, J. L., Sharma, P., Tokas, I., Pandey, A., Luan, S., & Pandey, G. K. (2018). Protein phosphatase 2C, AP2C1 interacts with and negatively regulates the function of CIPK9 under potassium deficient conditions in Arabidopsis. *Journal of Experimental Botany.*

Srivastava, A. K., Shankar, A., Chandran, A.K., Sharma, M., Jung, K.H., Suprasanna, P., and Pandey, G.K. (2019). Emerging concepts of potassium homeostasis in plants. *Journal of Experimental Botany, 71*(2), 608–619.

Tiwari, K. N. (2002). Nutrient deficiency symptoms in rice. In Better Crops Int. (Saskatchewan, Canada.), pp. 23–25.

Very, A. A., & Davies, J. M. (2000). Hyperpolarization-activated calcium channels at the tip of Arabidopsis root hairs. *Proceedings of the National Academy of Sciences of the United States of America, 97*, 9801–9806.

Vicente-Agullo, F., Rigas, S., Desbrosses, G., Dolan, L., Hatzopoulos, P., & Grabov, A. (2004). Potassium carrier TRH1 is required for auxin transport in Arabidopsis roots. *The Plant Journal, 40*, 523–535.

Walker, D. J., Leigh, R. A., & Miller, A. J. (1996). Potassium homeostasis in vacuolate plant cells *Proceedings of the National Academy of Sciences of the United States of America, 93*, 10510–10514.

Wang, Y., & Wu, W. H. (2010). Plant sensing and signaling in response to K+-deficiency. *Molecular Plant, 3*, 280–287.

Wang, Y., & Wu, W. H. (2013). Potassium transport and signaling in higher plants. *Annual Review of Plant Biology, 64*, 451–476.

Wang, T. B., Gassmann, W., Rubio, F., Schroeder, J. I., & Glass, A. D. (1998). Rapid up-regulation of HKT1, a high-affinity potassium transporter gene, in roots of barley and wheat following withdrawal of potassium1. *Plant Physiology, 118*, 651–659.

Wang, S. S., Diao, W. Z., Yang, X., Qiao, Z., Wang, M., Acharya, B. R., & Zhang, W. (2015). Arabidopsis thaliana CML25 mediates the Ca(2+) regulation of K(+) transmembrane trafficking during pollen germination and tube elongation. *Plant, Cell & Environment, 38*, 2372–2386.

Wang, X., Hao, L., Zhu, B., & Jiang, Z. (2018). Plant calcium signaling in response to potassium deficiency. *International Journal of Molecular Sciences, 19*.

Wu, X., Qiao, Z., Liu, H., Acharya, B. R., Li, C., & Zhang, W. (2017). CML20, an Arabidopsis calmodulin-like protein, negatively regulates guard cell ABA signaling and drought stress tolerance. *Frontiers in Plant Science, 8*, 824.

Xiao-Li, T., Gang-Wei, W., Rui, Z., Pei-Zhu, Y., Liu-Sheng, D., & Zhao-Hu, L. (2008). Conditions and indicators for screening cotton (Gossypium hirsutum L.) varieties tolerant to low potassium. *Acta Agronomica Sinica, 34*, 1435–1443.

Xu, J., Li, H. D., Chen, L. Q., Wang, Y., Liu, L. L., He, L., & Wu, W. H. (2006). A protein kinase, interacting with two calcineurin B-like proteins, regulates K+ transporter AKT1 in Arabidopsis. *Cell, 125*, 1347–1360.

Yang, T., & Poovaiah, B. W. (2002). Hydrogen peroxide homeostasis: Activation of plant catalase by calcium/calmodulin. *Proceedings of the National Academy of Sciences of the United States of America, 99*, 4097–4102.

Yang, X. E., Liu, J. X., Wang, W. M., Li, H., Luo, A. C., Ye, Z. Q., & Yang. (2003). Genotypic differences and some associated plant traits in potassium internal use efficiency of lowland rice (Oryza sativa L.) | SpringerLink. *Nutrient Cycling in Agroecosystems, 67*, 273–282.

Zhao, M., Oosterhuis, D. M., & Bednarz, C. W. (2001). Influence of potassium deficiency on photosynthesis, chlorophyll content, and chloroplast ultrastructure of cotton plants. *Photosynthetica, 39*, 103–109.

Zhao, L. N., Shen, L. K., Zhang, W. Z., Zhang, W., Wang, Y., & Wu, W. H. (2013). Ca2+-dependent protein kinase11 and 24 modulate the activity of the inward rectifying K+ channels in Arabidopsis pollen tubes. *Plant Cell, 25*, 649–661.

Zou, J. J., Wei, F. J., Wang, C., Wu, J. J., Ratnasekera, D., Liu, W. X., & Wu, W. H. (2010). Arabidopsis calcium-dependent protein kinase CPK10 functions in abscisic acid- and Ca2+-mediated stomatal regulation in response to drought stress. *Plant Physiology, 154*, 1232–1243.

Chapter 8
Potassium Perception and Sensing

Introduction

The most prevalent way of absorbing minerals in plant cell is through roots. Epidermal cell and root hairs are mainly responsible for absorption of K^+ from the soil; hence, these are qualified as potential organs for K^+ sensing. Depending on the external K^+ concentration, dual-affinity mechanism for K^+ uptake operated at different external concentrations. And this suggests that higher plant seems to perceive $[K^+]_{ext}$ resulting in 'switching on' of the appropriate mechanism for efficient K^+ uptake (Wang and Wu 2013).

Overview of Potassium Perception and Sensing

Embedded in the soil, root surface encounters varying K^+ concentrations depending on its availability in the soil. Usually, the typical concentration of K^+ around the root surface ranges from 0.1 to 1 mM, which is comparatively lower than the concentration of K^+ present inside the plants (Schroeder et al. 1994; Maathuis 2009). Therefore, the only choice for the root cells to absorb K^+ from the soil is that the absorption must be performed against K^+ concentration gradient. Along with constant perception of varying $[K^+]_{ext}$, root cells have developed mechanisms capable of K^+ absorption against the concentration gradient, utilizing impressive set of transport systems (Wang and Wu 2013). K uptake mediated by roots exhibits dual-affinity mechanism where two uptake mechanisms, viz. high-affinity uptake mechanism and low-affinity uptake mechanism, are governed by $[K^+]_{ext}$ concentration in the soil. A low $[K^+]_{ext}$ concentration below 0.2 mM leads to the activation of high-affinity K^+ uptake mechanisms mediated majorly by K^+ transporters, and a relatively high $[K^+]_{ext}$ above 0.3 mM leads to the activation of low-affinity K^+ uptake

© The Author(s), under exclusive license to Springer Nature Switzerland AG 2020
G. K. Pandey, S. Mahiwal, *Role of Potassium in Plants*, SpringerBriefs in Plant Science, https://doi.org/10.1007/978-3-030-45953-6_8

mediated by channels (Epstein et al. 1963; Maathuis and Sanders 1994; Schroeder et al. 1994). At high $[K^+]_{ext,}$ K^+ simply takes a path through channels to move down the electrochemical gradient indicating efficient utilization of the channel function, whereas at low $[K^+]_{ext}$ in order to trail K^+ inside the cells against the electrochemical gradient, active transporter systems are required justifying the role of transporters in high-affinity uptake (Ragel et al. 2019). This attribute of sensing $[K^+]_{ext}$ concentration by the root cell plasma membrane gives identity to the root cell as a potent K^+ sensor. Root cells have been implicated as a promising candidate component involved in low K^+ sensing and deficiency signaling (Wang and Wu 2013). However, even with the availability of advanced tools and techniques, no clue has been provided on the molecular component(s)/mechanism related to the K^+ perception and sensing in plants. Present knowledge in this field needs further investigations on this long-standing open question.

Mechanism of Potassium Perception and Sensing in Plants

Once the K deficiency is perceived at plasma membrane of root epidermal cell, its subsequent transduction to the cytoplasm leads to a series of physiological and biochemical reactions that may generate either short- or long-term responses (Wang and Wu 2013). The short-term responses may provide an insight into molecular mechanism involved in K perception as the sensing mechanisms prefer to generate a 'quick response' instead of delayed response (Wang and Wu 2013). Several hypotheses have been proposed about the sensing mechanisms; out of those the most promising ones have been discussed in this chapter.

Hypothesis 1: Role of Membrane Potential in K⁺ Sensing

External K^+ concentration influences the membrane potential although the crucial job of membrane potential maintenance is primarily mediated by H^+-ATPase activity (Maathuis and Sanders 1993; Palmgren 2001). The root cell membrane exhibits high specificity and dependency on $[K^+]_{ext}$, hence displaying linear relationship with external K^+ concentration (Maathuis and Sanders 1993). The varying $[K]_{ext}$ concentration affects the membrane potential in such a way that low $[K^+]_{ext}$ leads to hyperpolarization of the membrane, whereas high $[K^+]_{ext}$ leads to depolarization of the membrane (Schroeder and Fang 1991; Maathuis and Sanders 1993; Spalding et al. 1999; Nieves-Cordones et al. 2008). The change in membrane potential is thought to serve as a sensing signal in response to K^+ deprivation in plant cells due to its appearance after few minutes of K^+ starvation (Fig. 8.1) (Maathuis and Sanders 1993; Nieves-Cordones et al. 2008). However, the mechanism underlying membrane potential triggered specific responses need to be deciphered.

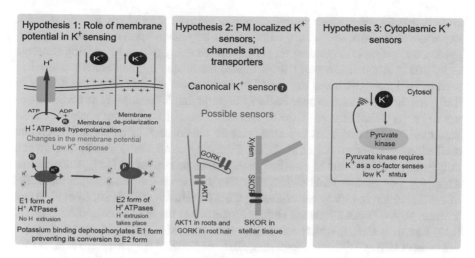

Fig. 8.1 Hypothesis related to the K⁺ sensing in plant. Hypothesis 1 supports the view that change in membrane potential is involved in low K⁺ sensing response. Hypothesis 2 supports the view that PM-localized channels and transporters act as possible sensors in root as well as stellar tissue. Hypothesis 3 states that cytosolic enzyme such as pyruvate kinases acts as a cytoplasmic sensor

A well-known H-ATPase also exhibits sensitivity to external K. This H⁺-ATPase has two forms, E1 and E2, which are interconvertible into each other depending on phosphorylation status. E1 form is ATP-hydrolysing form, which upon phosphorylation gets converted into E2 form. It is hypothesized that E1 form of the ATPases (E1P) (ATP-hydrolysing form) is dephosphorylated upon K⁺ binding. This dephosphorylation event takes place on an aspartate residue in the P-domain, preventing its conversion to the E2 form (E2P). The overall effect of the process can be seen as an obstruction to H⁺ extrusion, hence uncoupling ATP hydrolysis and H⁺ pumping activity. In this study, K⁺ has been suggested as an intrinsic uncoupler of the proton pump, which may play a role in regulating H⁺/ATP coupling ratio. This K⁺-induced dephosphorylation event negatively regulates transmembrane electrochemical gradient when E1P form of the ATPase enzyme is accumulated in the cell (Fig. 8.1) (Buch-Pedersen et al. 2006; Cherel et al. 2014). Physiological data supporting these hypotheses is still lacking; experimental support is required to establish the role of pyruvate kinases and H⁺-ATPases in K⁺ sensing and signaling.

Hypothesis 2: Plasma Membrane-Localized K⁺ Sensors

The inherent capacity of K⁺ sensing is exhibited by plasma membrane of root epidermal cell making them the possible sites where K⁺ sensors could be located (Wang and Wu 2013). Interestingly, regardless of extensive information available, no canonical K⁺ sensor could be identified so far. Previous reports indicate that when [K⁺]$_{ext}$ is reduced, outward-rectifying K⁺ channels may function as K⁺ sensors.

GORK and SKOR are two hypothesized as potential candidates for this. GORK expresses in the root hair where its activity is dependent upon membrane depolarization and gating is dependent on $[K^+]_{ext}$ thus can play role in K^+ sensing (Sharma et al. 2013). Another channel SKOR is also modulated by membrane potential as well as $[K^+]_{ext.}$ SKOR being expressed in stellar tissue would control the K^+ flux to the xylem in accordance to apoplastic K^+ highlighting K^+ sensitivity of SKOR channel to $[K^+]_{ext}$ (Johansson et al. 2006; Sharma et al. 2013). Another potential candidate could be the major inward-rectifying K^+ channel AKT1 as it exhibits typical dual-affinity character (Li et al. 2017) (Spalding et al. 1999). AKT1 can switch between low-affinity and high-affinity range upon perception of $[K^+]_{ext}$ fluctuations. This $[K^+]_{ext}$-dependent kinetic shift may also depend on the phosphorylation status of AKT1, which in turn is controlled by CIPK23 (Hirsch et al. 1998; Li et al. 2006; Yuan et al. 2007). AKT1 shows similar properties as the nitrate transporter NRT1.1 (also known as CHL1, which acts as nitrate sensor); also, both are regulated by the same set of CBL-CIPK complexes, i.e., CBL1/9-CIPK23 (Fig. 8.1) (Ho et al. 2009). All these facts have led to the assumption that AKT1 may act as K^+ sensor for its activation via CBL-CIPK complexes by triggering their mobilization.

Hypothesis 3: Cytoplasmic K^+ Sensors

A short-term response to K deficiency is mobilization of K^+ from the vacuole to the cytosol in order to compensate for the decreased K^+ absorption from external media. In longer term, cytosolic activity is known to be decreased after K^+ starvation affecting many physiological processes in plant cells (Walker et al. 1996). The hypothesis suggests that cytosolic K^+ concentration can be sensed by two means: either by K^+-sensitive cytosolic enzyme or by plasma membrane proton pumps. Some of the cytosolic enzymes require K^+ as a co-factor; among all those, the most likely candidate for sensing K^+ concentration is enzyme pyruvate kinase as it shows maximum sensitivity to cytosolic K^+ concentration (Ramirez-Silva et al. 2001; Amtmann and Armengaud 2009; Armengaud et al. 2009). If plants are subjected to long-term K deficiency, inhibition of pyruvate kinase activity can be observed in root cells (Fig. 8.1) (Armengaud et al. 2009). Interestingly, two separate experiments reported contrasting observations; based on the experimental conditions, pyruvate kinase activity was found to be greatly enhanced in leaves of wheat plant and reduced in *Arabidopsis* roots. However, these contrasting observations may also arise from the different mechanisms of K^+ deficiency in two plant species as the K^+ requirements of both the plants significantly vary (Sugiyama et al. 1968; Armengaud et al. 2009).

Plants need to ensure continuous availability of soil K^+; in order to fulfil this requirement, they have devised complex mechanisms that can sense K^+ around the root surface. The entire mechanism is not deciphered yet. Few hypotheses have been discussed in this chapter based on the present-day understanding of K^+ perception in the roots. The so-called K^+ sensory system has not been identified in plants so far.

References

Amtmann, A., & Armengaud, P. (2009). Effects of N, P, K and S on metabolism: New knowledge gained from multi-level analysis. *Current Opinion in Plant Biology, 12,* 275–283.

Armengaud, P., Sulpice, R., Miller, A. J., Stitt, M., Amtmann, A., & Gibon, Y. (2009). Multilevel analysis of primary metabolism provides new insights into the role of potassium nutrition for glycolysis and nitrogen assimilation in Arabidopsis roots. *Plant Physiology, 150,* 772–785.

Buch-Pedersen, M. J., Rudashevskaya, E. L., Berner, T. S., Venema, K., & Palmgren, M. G. (2006). Potassium as an intrinsic uncoupler of the plasma membrane H+-ATPase. *The Journal of Biological Chemistry, 281,* 38285–38292.

Cherel, I., Lefoulon, C., Boeglin, M., & Sentenac, H. (2014). Molecular mechanisms involved in plant adaptation to low K(+) availability. *Journal of Experimental Botany, 65,* 833–848.

Epstein, E., Rains, D. W., & Elzam, O. E. (1963). Resolution of dual mechanisms of potassium absorption by barley roots. *Proceedings of the National Academy of Sciences of the United States of America, 49,* 684–692.

Hirsch, R. E., Lewis, B. D., Spalding, E. P., & Sussman, M. R. (1998). A role for the AKT1 potassium channel in plant nutrition. *Science, 280,* 918–921.

Ho, C. H., Lin, S. H., Hu, H. C., & Tsay, Y. F. (2009). CHL1 functions as a nitrate sensor in plants. *Cell, 138,* 1184–1194.

Johansson, I., Wulfetange, K., Poree, F., Michard, E., Gajdanowicz, P., Lacombe, B., Sentenac, H., Thibaud, J. B., Mueller-Roeber, B., Blatt, M. R., & Dreyer, I. (2006). External K+ modulates the activity of the Arabidopsis potassium channel SKOR via an unusual mechanism. *The Plant Journal, 46,* 269–281.

Li, L., Kim, B. G., Cheong, Y. H., Pandey, G. K., & Luan, S. (2006). A Ca2+ signaling pathway regulates a K+ channel for low-K response in Arabidopsis. *Proceedings of the National Academy of Sciences of the United States of America, 103,* 12625–12630.

Li, J., Wu, W. H., & Wang, Y. (2017). Potassium channel AKT1 is involved in the auxin-mediated root growth inhibition in Arabidopsis response to low K(+) stress. *Journal of Integrative Plant Biology, 59,* 895–909.

Maathuis, F. J. (2009). Physiological functions of mineral macronutrients. *Current Opinion in Plant Biology, 12,* 250–258.

Maathuis, F. J. M., & Sanders, D. (1993). Energization of potassium uptake in Arabidopsis thaliana | SpringerLink. *Planta, 191,* 302–307.

Maathuis, F. J., & Sanders, D. (1994). Mechanism of high-affinity potassium uptake in roots of Arabidopsis thaliana *Proceedings of the National Academy of Sciences of the United States of America, 91,* 9272–9276.

Nieves-Cordones, M., Miller, A. J., Aleman, F., Martinez, V., & Rubio, F. (2008). A putative role for the plasma membrane potential in the control of the expression of the gene encoding the tomato high-affinity potassium transporter HAK5. *Plant Molecular Biology, 68,* 521–532.

Palmgren, M. G. (2001). Plant plasma membrane H+-ATPases: Powerhouses for nutrient uptake. *Annual Review of Plant Physiology and Plant Molecular Biology, 52,* 817–845.

Ragel, P., Raddatz, N., Leidi, E. O., Quintero, F. J., & Pardo, J. M. (2019). Regulation of K+ nutrition in plants. *Frontiers in Plant Science, 10,* 281.

Ramirez-Silva, L., Ferreira, S. T., Nowak, T., Tuena de Gomez-Puyou, M., & Gomez-Puyou, A. (2001). Dimethylsulfoxide promotes K+-independent activity of pyruvate kinase and the acquisition of the active catalytic conformation. *European Journal of Biochemistry, 268,* 3267–3274.

Schroeder, J. I., & Fang, H. H. (1991). Inward-rectifying K+ channels in guard cells provide a mechanism for low-affinity K+ uptake. *Proceedings of the National Academy of Sciences of the United States of America, 88,* 11583–11587.

Schroeder, J. I., Ward, J. M., & Gassmann, W. (1994). Perspectives on the physiology and structure of inward-rectifying K+ channels in higher plants: Biophysical implications for K+ uptake. *Annual Review of Biophysics and Biomolecular Structure, 23,* 441–471.

Sharma, T., Dreyer, I., & Riedelsberger, J. (2013). The role of K+ channels in uptake and redistribution of potassium in the model plant Arabidopsis thaliana. *Frontiers in Plant Science, 4,* 224.

Spalding, E. P., Hirsch, R. E., Lewis, D. R., Qi, Z., Sussman, M. R., & Lewis, B. D. (1999). Potassium uptake supporting plant growth in the absence of AKT1 channel activity: Inhibition by ammonium and stimulation by sodium. *The Journal of General Physiology, 113,* 909–918.

Sugiyama, T., Goto, Y., & Akazawa, T. (1968). Pyruvate kinase activity of wheat plants grown under potassium deficient conditions. *Plant Physiology, 43,* 730–734.

Walker, D. J., Leigh, R. A., & Miller, A. J. (1996). Potassium homeostasis in vacuolate plant cells. *Proceedings of the National Academy of Sciences of the United States of America, 93,* 10510–10514.

Wang, Y., & Wu, W. H. (2013). Potassium transport and signaling in higher plants. *Annual Review of Plant Biology, 64,* 451–476.

Yuan, L., Loque, D., Kojima, S., Rauch, S., Ishiyama, K., Inoue, E., Takahashi, H., & von Wiren, N. (2007). The organization of high-affinity ammonium uptake in Arabidopsis roots depends on the spatial arrangement and biochemical properties of AMT1-type transporters. *Plant Cell, 19,* 2636–2652.

Chapter 9
Emerging Roles of Potassium in Plants

Introduction

Prominently known for its nutritional value since few decades, K is well implicated in plant growth, development and stress tolerance. Apart from this, there may be a grey area which is yet to be explored. Recently, few reports discuss about the possible role of K as a signaling molecule based on the several lines of indication. For optimal operation of plant metabolic machinery, maintenance of intracellular K is a prerequisite which gets affected by various biotic and abiotic stresses ultimately resulting in reduction of crop quality. In the context of abiotic stress, role of K has been studied chasmicly with lots of available literature, whereas its role in biotic stress is still emerging. Reports highlighting its crucial role in defence against biotic stress have started to come up. Furthermore, due to lack of direct experimental evidences of the same, more and more experimental data need to be generated in order to establish new roles of K in plants.

Potassium in Biotic Stress

Biotic factors are known to reduce crop quality and production since centuries. In agronomy, with the modern farming technologies as well as available tools and techniques, several methods are employed to evade biotic stress agents away from the crops, effectively resulting in qualitative as well as quantitative crop production. Although these methods seem to be promising, they may not be effective every time against each pathogen attack. So the only propitious option left for the plant is their strong defence system. The plant defence system is known to be influenced by the nutritional status of the plant. Nutritional status of K poses a strong impact on the plant vulnerability to pathogens (Wang and Wu 2013). An earlier report suggests

© The Author(s), under exclusive license to Springer Nature Switzerland AG 2020
G. K. Pandey, S. Mahiwal, *Role of Potassium in Plants*, SpringerBriefs in Plant
Science, https://doi.org/10.1007/978-3-030-45953-6_9

that the K deficiency in plants increases the pathogen susceptibility and the chances of infection as compared to K-sufficient plants. This has been observed in the case of rice borer infestation where the infection was substantially present in absence of K^+ but showed a rapid reduction upon increasing K^+ concentration (Sarwar 2012). Similar mechanistic influences were observed in crop *Cornus florida L.* upon infection with *Discula destructiva* redline (Holzmueller et al. 2007). Higher K content may reduce internal nutritional competition among pathogens (Holzmueller et al. 2007). If sufficient amount of K^+ is available, plant utilizes more nutrients for plant defence, protection and damage repair along with allocation of more nutrients to form stronger cell wall, which eventually protects against pathogen attack (Mengel and Kirkby 2001). Other preventive mechanisms against invading pathogens are also reported such as increased strength of culm and stalk in rice and stomata closure during airborne pathogen infections in presence of adequate K^+ supply (Wang and Wu 2013).

Also, disease incidence of stem rot and aggregate sheath spots were significantly reduced upon supplementation with K fertilizer, and there was negative correlation among disease severity and percentage of K in leaf blades (Williams and Smith 2001). Interestingly, application of K fertilizer does not always reduce the chances of pathogen infection; sometimes it may stimulate an opposite effect or no effect at all. In strawberry plant supplementation with excess K^+ increased the chances of infection by anthracnose pathogen (*Colletotrichum gloeosporioides*), whereas enhanced resistance was observed when no K^+ was supplemented (Nam et al. 2006). This could be because of synthesis of phytohormones such as auxin, ethylene and jasmonic acid and molecules such as reactive oxygen species (ROS) induced by low K^+ status of the plant and resulted in increased stress tolerance (Ashley et al. 2006; Amtmann et al. 2008). Amount and source of K^+ as well as pathogen species may be a deciding factor for resistance or susceptibility of the plant (Wang and Wu 2013).

K homeostasis and transport system is also known to play an important role in virus-host recognition. Changes in the K^+ fluxes just after 10 min. of viral inoculation were reported to be helpful in identifying early stage of viral infection (Shabala et al. 2010). Further pharmacological and membrane potential measurement experiments suggest involvement of K outward-rectifying (KOR) channels in regulating a part of these sudden K^+ fluxes. These channels are supposed to be originated from incorporation of viral K^+ efflux channels into host plasma membrane probably to reduce cell turgor pressure assisting the entry of viral DNA into the host cell (Neupärtl et al. 2008; Shabala and Pottosin 2014). Overall, K status determines the pathogen susceptibility and disease resistance in plants.

Can Potassium Act as a Signaling Molecule?

A common feature for the stress tolerance in plant seems to be the ability of the plants to retain K^+ as seen in most of the cases, but recent reports put emphasis on importance of stress-induced K^+ efflux in regulating growth and development under unfavourable conditions. Considering the vitality of K retention for stress tolerance, the need of developing K$^+$ efflux channels in plants is questionable (Shabala 2017). A totally different physiological mechanism may be functioning in plant that is way beyond our imagination. Many hypotheses have been given to explain all these phenomena, and many researchers speculate a signaling role of K$^+$ in plant growth during hostile conditions. Under salinity stress, K$^+$ efflux is thought as a 'safety valve' to counter with initial membrane depolarization, till upregulated H$^+$-ATPases gradually compensate to restore the membrane potential which is again followed by K$^+$ retention by high-affinity transporters present in the plants (Jinglan and Seliskar 1998; Chen et al. 2007; Alvarez-Pizarro et al. 2009). Under salinity and oxidative stress, high K$^+$ retention is required to inhibit the activity of caspase-like proteases and endonucleases to minimize damage induced by programmed cell death (PCD). This is supported by the experimental evidence showing null mutant of *Arabidopsis* GORK channel exhibit less PCD compared to wild-type plants (Demidchik et al. 2010). Another hypothesis also termed as 'metabolic hypothesis' suggests that K$^+$ efflux may act as a metabolic switch responsible for conserving energy for repair and adaptation and obstructing the energy-consuming anabolic reactions (Fig. 9.1) (Demidchik 2014). During stress, plants must maintain the balance between metabolism and defence along with allocation of more ATP pools to the plant defence. In order to avoid competition among the two processes, cell metabolism is shut down so that more energy can be contributed to defence responses. This is achieved by

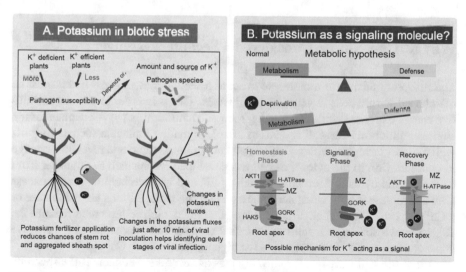

Fig. 9.1 (**a**) K in biotic stress; pathogen susceptibility depends on the amount and source of K and pathogen species regulating disease responses in plants. (**b**) Role of K$^+$ as a signaling molecule: metabolic hypothesis states that under normal conditions, a balance between metabolism and defence responses exists, whereas in K deficiency, metabolism is shut down to a level diverting major energy to defence responses. A model for K$^+$ signaling has been suggested, which consists of the homeostasis phase, signaling phase and final recovery phase. (Adapted and modified from Shabala 2017)

decreasing cytosolic K^+ to a sub-threshold level and redistributing ATP pools towards defence responses by inactivating most metabolic reactions (Shabala 2017). A hypothetical mechanism has been suggested for cytosolic K^+ signaling which consist of three phases (Fig. 9.1), the first phase being the homeostasis or recovery phase, the second being the signaling phase and the last being the final recovery phase. It is assumed that homeostasis phase is characterized by slightly higher cytosolic K^+ concentration in the root apex zone as compared to root maturation zone because of high demand for K^+ (Chen et al. 2005). In maturation zone, AKT channels maintain constant cytosolic K^+ levels, whereas H^+-ATPases provide electrical gradients for membrane potential (MP) maintenance, and in apex zone, less negative MP results in K^+ leak via GORK, which is compensated by HAK family transporters (Chakraborty et al. 2016). Signaling phase is characterized by depolarization of MP to very low levels in the apex triggering tremendous K^+ efflux through GORK channels upon perception of stress. This K^+ efflux acts as a switch and is responsible for restoring the MP as well as initiates defence operation of the cell. The recovery phase is characterized by recovery of K^+ in both zones via stress-induced activation of H^+-ATPases along with symplastic K^+ contribution from the mature zones to the apex. All these hypotheses need experimental validations to establish K^+ as a signaling molecule. It is still debatable whether K^+ can be stated as potential signaling molecule or not. Many questions need to be answered, and many mechanisms need to be deciphered with experimental support to add K^+ as another second messenger in the near future (Shabala 2017).

Why K^+ Can Not Be Considered a Second Messenger?

The first prerequisite for a molecule to act as a second messenger is that it must be maintained within cells at very low basal levels. This is in order to prevent prohibitive energy expenses while modulating their concentration as per the demand of the signaling function (Carafoli and Krebs 2016). Signaling molecule such as Ca^{2+} fits well into this criterion, where Ca^{2+} is well known for its rapid and reversible change in concentration in the cytosol as cell has developed mechanism to sequester extra Ca^{2+} from the cytosol. Storehouses such as ER and vacuole help in storing excess Ca^{2+} and release it only when needed, in order to generate either a Ca^{2+} signature or a Ca^{2+} response (Stael et al. 2012). On the other hand, high cellular requirements for K^+ in the cytosol make it unfit to be called as a second messenger as the relative concentration of K^+ in the cytosol is too high as compared to Ca^{2+}.

One of the main reasons for utilizing ions as second messengers is the speed by which they can be generated without any enzymatic steps. The speed of the response is dependent upon the intracellular ion concentration and proximity of the ion to its targets (Newton et al. 2016). Ca^{2+} is well known for generating quick responses in milliseconds (Berridge 2006), whereas for K^+ no such distinctions have been made so far. It is not clear how rapidly K^+ fluxes occur in a cell that may give rise to a response.

To conclude this chapter, K+ is a versatile component in plant nutrition regulating several aspects such as plant development and abiotic as well as biotic stresses. Recently, researchers are hypothesizing its role in signaling and hence suggesting it as a second messenger. The concept needs further support in order to establish this as a signaling molecule.

References

Alvarez-Pizarro, J. C., Gomes-Filho, E., de Lacerda, C. F., Alencar, N. L. M., & Prisco, J. T. (2009). Salt-induced changes on H+-ATPase activity, sterol and phospholipid content and lipid peroxidation of root plasma membrane from dwarf-cashew (Anacardium occidentale L.) seedlings I SpringerLink. *Plant Growth Regulation, 59*, 125–135.

Amtmann, A., Troufflard, S., & Armengaud, P. (2008). The effect of potassium nutrition on pest and disease resistance in plants. *Physiologia Plantarum, 133*, 682–691.

Ashley, M. K., Grant, M., & Grabov, A. (2006). Plant responses to potassium deficiencies: A role for potassium transport proteins. *Journal of Experimental Botany, 57*, 425–436.

Berridge, M. J. (2006). Calcium microdomains: Organization and functions. *Cell Calcium, 40*(5–6), 405–412.

Carafoli, E., & Krebs, J. (2016). Why Calcium? How Calcium became the best communicator. *The Journal of Biological Chemistry, 29*, 20849–20857.

Chakraborty, K., Bose, J., Shabala, L., Eyles, A., & Shabala, S. (2016). Evaluating relative contribution of osmotolerance and tissue tolerance mechanisms toward salinity stress tolerance in three Brassica species. *Physiologia Plantarum, 158*, 135–151.

Chen, Z., Newman, I., Zhou, M., Mendham, N., Zhang, G., & Shabala, S. (2005). Screening plants for salt tolerance by measuring K flux: A case study for barley. *Plant, Cell & Environment, 28*, 1230–1246.

Chen, Z., Pottosin, I. I., Cuin, T. A., Fuglsang, A. T., Tester, M., Jha, D., Zepeda-Jazo, I., Zhou, M., Palmgren, M. G., Newman, I. A., & Shabala, S. (2007). Root plasma membrane transporters controlling K+/Na+ homeostasis in salt-stressed barley. *Plant Physiology, 145*, 1714–1725.

Demidchik, V. (2014). Mechanisms and physiological roles of K+ efflux from root cells. *Journal of Plant Physiology, 171*(9), 696–707.

Demidchik, V., Cuin, T. A., Svistunenko, D., Smith, S. J., Miller, A. J., Shabala, S., Sokolik, A., & Yurin, V. (2010). Arabidopsis root K+-efflux conductance activated by hydroxyl radicals: Single-channel properties, genetic basis and involvement in stress-induced cell death. *Journal of Cell Science, 123*, 1468–1479.

Holzmueller, E. J., Jose, S., & Jenkins, M. A. (2007). Influence of calcium, potassium, and magnesium on Cornus florida L. density and resistance to dogwood anthracnose I SpringerLink. *Plant and Soil, 290*, 189–199.

Jinglan, W., & Seliskar, D. M. (1998). Salinity adaptation of plasma membrane H+-ATPase in the salt marsh plant Spartina patens: ATP hydrolysis and enzyme kinetics. *Journal of Experimental Botany, 49*, 1005–1013.

Mengel, K., & Kirkby, E. A. (2001). *Principles of plant nutrition*. Dordrecht: Springer.

Nam, M. H., Jeong, S. K., Lee, Y. S., Choi, J. M., & Kim, H. G. (2006). Effects of nitrogen, phosphorus, potassium and calcium nutrition on strawberry anthracnose. *Plant Pathology, 55*, 246–249.

Neupärtl, M., Meyer, C., Woll, I., Frohns, F., Kang, M., Van Etten, J.L., Kramer, D., Hertel, B., Moroni, ., Thiel, G. (2008). Chlorella viruses evoke a rapid release of K+ from host cells during the early phase of infection. *Virology, 372*(2), 340–348.

Newton, A. C., Bootman, M. D., & Scott, J. D. (2016). Second messengers. *Cold Spring Harbor Perspectives in Biology, 8*(8).

Sarwar, M. H. (2012). Effects of potassium fertilization on population build up of rice stem borers (lepidopteron pests) and rice (Oryza sativa L.) yield. *Journal of Cereals and Oilseeds, 3*, 6–9.

Shabala, S. (2017). Signalling by potassium: Another second messenger to add to the list? *Journal of Experimental Botany, 68*, 4003–4007.

Shabala, S., & Pottosin, I. (2014). Regulation of potassium transport in plants under hostile conditions: Implications for abiotic and biotic stress tolerance. *Physiologia Plantarum, 151*, 257–279.

Shabala, S., Babourina, O., Rengel, Z., & Nemchinov, L. G. (2010). Non-invasive microelectrode potassium flux measurements as a potential tool for early recognition of virus-host compatibility in plants. *Planta, 232*, 807–815.

Stael, S., Wurzinger, B., Mair, A., Mehlmer, N., Vothknecht, U. C., & Teige, M. (2012). Plant organellar calcium signaling: An emerging field. *Journal of Experimental Botany, 63*(4), 1525–1542.

Wang, Y., & Wu, W. H. (2013). Potassium transport and signaling in higher plants. *Annual Review of Plant Biology, 64*, 451–476.

Williams, J., & Smith, S. G. (2001). Correcting potassium deficiency can reduce rice stem diseases. *Better Crops, 85*, 7–9.

Chapter 10
Key Questions and Future Perspective

The importance of K as a nutrient in all forms of life has been realized decades ago, and huge amount of efforts have been invested in the past to understand its role in plant as well as in animal. Properly documented for its role in animal and plant, the conventional requirement for any cell is to maintain K homeostasis. In each cell, starting from prokaryotes to eukaryotes, several mechanisms do exist to keep a check on K homeostasis. One thing that is common among lower and higher organism is that they all possess same 'decorative designs', termed as channels and transporters, to maintain K homeostasis (Miller 2000). These transporter and channels differ in their activity, selectivity and regulation, giving rise to multitude of responses in a cell. The intracellular abundance of K in a cell is itself an indicator of its immense significance in the biological system (Dreyer and Uozumi 2011).

In the recent years, extraordinary progress has been made in the field of K research. With available genome sequence information from many organisms such as *Escherichia coli* (Blattner et al. 1997), *Saccharomyces cerevisiae, Caenorhabditis elegans, Mus musculus* (Waterston et al. 2002), *Homo sapiens* (Venter et al. 2001), *Arabidopsis thaliana*, etc., K uptake systems have been well identified and characterized. Solid experimental designs utilizing cutting-edge modern technologies are the major driving force for the fast-track progress in recent years flourishing K research. Although the available information is sufficient to build a conceptual basis for role and importance of K, there are many doors ahead waiting to be unlocked.

K holds colossal eminence in agriculture and plant biology. One of the major challenges faced by a plant is to make use of K resources from the soil minerals, which seems like a tip of the iceberg. As the readily available pools of K for the plants are limited, understanding the process of K release from the soil minerals is very crucial in order to make them accessible for the plant. New approaches need to be developed to extract the K trapped in the soil minerals so that plants do not encounter K deficiency in the soil. Another area of focus for the plant biologists is improvisation of K acquisition and K utilization efficiency via genetic modifications in the plants (Srivastava et al. 2019). Several attempts for this have been made

© The Author(s), under exclusive license to Springer Nature Switzerland AG 2020

G. K. Pandey, S. Mahiwal, *Role of Potassium in Plants*, SpringerBriefs in Plant Science, https://doi.org/10.1007/978-3-030-45953-6_10

in various crop species. Plant biologists are in constant strive as this may open the sky of new opportunities for sustainable agriculture production. Incessant attempts are being made to generate plants that can tolerate or sustain under low K conditions. The agriculturist has deciphered the symptoms of K deficiency and their remedies extensively. The plant biologist had taken this to another level, where most of the plant responses under K deficiency are well studied and documented. Pathways regulating various channels and transporters in harmony with the signaling pathways have been decrypted (Pandey et al. 2007; Ragel et al. 2015). However, our limited knowledge of the underlying molecular mechanism involved in sensing K deficiency in plants is a rugged barrier to establish the fundamentals of K sensing and response in plants. Nevertheless, one of the interesting aspects of K has been studied well in due course, which tends to suggest a signaling role for K.

Increasing evidences are suggesting a role for nitrate transporter1 (NRT1) which can function as a proton-coupled K^+/H^+ antiporter (Li et al. 2017). It is important to note that in plants K^+/NO_3^- absorption and transport are somehow known to be coordinated, further providing an indication for deciphering this coordination. Engineering these associated transporters such as NRT1 as well can possibly increase K^+ uptake efficiency. Since plant nutrition depends on the complex interplay of all major and minor nutrients, it would be interesting to determine whether any other components of another nutrient pathway interact with K^+ pathway. As we have looked into broader role of K^+ in plants, abiotic and biotic stresses do affect plant at almost every stage; probably connecting the dots between various nutrients along with K can solve this. A long-standing question currently is whether K homeostasis in combination with other nutrients such as N and P can aid stress tolerance in plants.

There are speculations that together with well-known second messengers such as Ca^{2+} and ROS, K may serve as a signal that can configure stress-adaptive responses in plants. Experimental evidences are indeed required to substantiate these hypotheses in the near future that may expand the horizons of K biology.

References

Blattner, F. R., Plunkett, G., 3rd, Bloch, C. A., Perna, N. T., Burland, V., Riley, M., Collado-Vides, J., Glasner, J. D., Rode, C. K., Mayhew, G. F., Gregor, J., Davis, N. W., Kirkpatrick, H. A., Goeden, M. A., Rose, D. J., Mau, B., & Shao, Y. (1997). The complete genome sequence of Escherichia coli K-12. *Science, 277,* 1453–1462.

Dreyer, I., & Uozumi, N. (2011). Potassium channels in plant cells. *The FEBS Journal, 278,* 4293–4303.

Li, H., Yu, M., Du, X. Q., Wang, Z. F., Wu, W. H., Quintero, F. J., Jin, X. H., Li, H. D., & Wang, Y. (2017). NRT1.5/NPF7.3 functions as a proton-coupled H+/K+ Antiporter for K+ loading into the xylem in Arabidopsis. *The Plant Cell, 29*(8), 2016–2026.

Miller, C. (2000). An overview of the potassium channel family. *Genome Biology, 1.* Reviews0004.

Pandey, G. K., Cheong, Y. H., Kim, B. G., Grant, J. J., Li, L., & Luan, S. (2007). CIPK9: A calcium sensor-interacting protein kinase required for low-potassium tolerance in Arabidopsis. *Cell Research, 17,* 411–421.

Ragel, P., Rodenas, R., Garcia-Martin, E., Andres, Z., Villalta, I., Nieves-Cordones, M., Rivero, R. M., Martinez, V., Pardo, J. M., Quintero, F. J., & Rubio, F. (2015). The CBL-interacting protein kinase CIPK23 regulates HAK5-mediated high-affinity K+ uptake in Arabidopsis roots. *Plant Physiology, 169*, 2863–2873.

Srivastava, A. K., Shankar, A., Chandran, A. K., Sharma, M., Jung, K. H., Suprasanna, P., & Pandey, G. K. (2019). Emerging concepts of potassium homeostasis in plants. *Journal of Experimental Botany, 71*(2), 608–619.

Venter, J. C., Adams, M. D., Myers, E. W., Li, P. W., Mural, R. J., Sutton, G. G., Smith, H. O., Yandell, M., Evans, C. A., Holt, R. A., Gocayne, J. D., Amanatides, P., Ballew, R. M., Huson, D. H., Wortman, J. R., Zhang, Q., Kodira, C. D., Zheng, X. H., Chen, L., Skupski, M., Subramanian, G., Thomas, P. D., Zhang, J., Gabor Miklos, G. L., Nelson, C., Broder, S., Clark, A. G., Nadeau, J., McKusick, V. A., Zinder, N., Levine, A. J., Roberts, R. J., Simon, M., Slayman, C., Hunkapiller, M., Bolanos, R., Delcher, A., Dew, I., Fasulo, D., Flanigan, M., Florea, L., Halpern, A., Hannenhalli, S., Kravitz, S., Levy, S., Mobarry, C., Reinert, K., Remington, K., Abu-Threideh, J., Beasley, E., Biddick, K., Bonazzi, V., Brandon, R., Cargill, M., Chandramouliswaran, I., Charlab, R., Chaturvedi, K., Deng, Z., Di Francesco, V., Dunn, P., Eilbeck, K., Evangelista, C., Gabrielian, A. E., Gan, W., Ge, W., Gong, F., Gu, Z., Guan, P., Heiman, T. J., Higgins, M. E., Ji, R. R., Ke, Z., Ketchum, K. A., Lai, Z., Lei, Y., Li, Z., Li, J., Liang, Y., Lin, X., Lu, F., Merkulov, G. V., Milshina, N., Moore, H. M., Naik, A. K., Narayan, V. A., Neelam, B., Nusskern, D., Rusch, D. B., Salzberg, S., Shao, W., Shue, B., Sun, J., Wang, Z., Wang, A., Wang, X., Wang, J., Wei, M., Wides, R., Xiao, C., Yan, C., Yao, A., Ye, J., Zhan, M., Zhang, W., Zhang, H., Zhao, Q., Zheng, L., Zhong, F., Zhong, W., Zhu, S., Zhao, S., Gilbert, D., Baumhueter, S., Spier, G., Carter, C., Cravchik, A., Woodage, T., Ali, F., An, H., Awe, A., Baldwin, D., Baden, H., Barnstead, M., Barrow, I., Beeson, K., Busam, D., Carver, A., Center, A., Cheng, M. L., Curry, L., Danaher, S., Davenport, L., Desilets, R., Dietz, S., Dodson, K., Doup, L., Ferriera, S., Garg, N., Gluecksmann, A., Hart, B., Haynes, J., Haynes, C., Heiner, C., Hladun, S., Hostin, D., Houck, J., Howland, T., Ibegwam, C., Johnson, J., Kalush, F., Kline, I., Koduru, S., Love, A., Mann, F., May, D., McCawley, S., McIntosh, T., McMullen, I., Moy, M., Moy, L., Murphy, B., Nelson, K., Pfannkoch, C., Pratts, E., Puri, V., Qureshi, H., Reardon, M., Rodriguez, R., Rogers, Y. H., Romblad, D., Ruhfel, B., Scott, R., Sitter, C., Smallwood, M., Stewart, E., Strong, R., Suh, E., Thomas, R., Tint, N. N., Tse, S., Vech, C., Wang, G., Wetter, J., Williams, S., Williams, M., Windsor, S., Winn-Deen, E., Wolfe, K., Zaveri, J., Zaveri, K., Abril, J. F., Guigo, R., Campbell, M. J., Sjolander, K. V., Karlak, B., Kejariwal, A., Mi, H., Lazareva, B., Hatton, T., Narechania, A., Diemer, K., Muruganujan, A., Guo, N., Sato, S., Bafna, V., Istrail, S., Lippert, R., Schwartz, R., Walenz, B., Yooseph, S., Allen, D., Basu, A., Baxendale, J., Blick, L., Caminha, M., Carnes-Stine, J., Caulk, P., Chiang, Y. H., Coyne, M., Dahlke, C., Mays, A., Dombroski, M., Donnelly, M., Ely, D., Esparham, S., Fosler, C., Gire, H., Glanowski, S., Glasser, K., Glodek, A., Gorokhov, M., Graham, K., Gropman, B., Harris, M., Heil, J., Henderson, S., Hoover, J., Jennings, D., Jordan, C., Jordan, J., Kasha, J., Kagan, L., Kraft, C., Levitsky, A., Lewis, M., Liu, X., Lopez, J., Ma, D., Majoros, W., McDaniel, J., Murphy, S., Newman, M., Nguyen, T., Nguyen, N., Nodell, M., Pan, S., Peck, J., Peterson, M., Rowe, W., Sanders, R., Scott, J., Simpson, M., Smith, T., Sprague, A., Stockwell, T., Turner, R., Venter, E., Wang, M., Wen, M., Wu, D., Wu, M., Xia, A., Zandieh, A., & Zhu, X. (2001). The sequence of the human genome. *Science, 291*, 1304–1351.

Waterston, R. H., Lindblad-Toh, K., Birney, E., Rogers, J., Abril, J. F., Agarwal, P., Agarwala, R., Ainscough, R., Alexandersson, M., An, P., Antonarakis, S. E., Attwood, J., Baertsch, R., Bailey, J., Barlow, K., Beck, S., Berry, E., Birren, B., Bloom, T., Bork, P., Botcherby, M., Bray, N., Brent, M. R., Brown, D. G., Brown, S. D., Bult, C., Burton, J., Butler, J., Campbell, R. D., Carninci, P., Cawley, S., Chiaromonte, F., Chinwalla, A. T., Church, D. M., Clamp, M., Clee, C., Collins, F. S., Cook, L. L., Copley, R. R., Coulson, A., Couronne, O., Cuff, J., Curwen, V., Cutts, T., Daly, M., David, R., Davies, J., Delehaunty, K. D., Deri, J., Dermitzakis, E. T., Dewey, C., Dickens, N. J., Diekhans, M., Dodge, S., Dubchak, I., Dunn, D. M., Eddy, S. R., Elnitski, L., Emes, R. D., Eswara, P., Eyras, E., Felsenfeld, A., Fewell, G. A., Flicek,

P., Foley, K., Frankel, W. N., Fulton, L. A., Fulton, R. S., Furey, T. S., Gage, D., Gibbs, R. A., Glusman, G., Gnerre, S., Goldman, N., Goodstadt, L., Grafham, D., Graves, T. A., Green, E. D., Gregory, S., Guigo, R., Guyer, M., Hardison, R. C., Haussler, D., Hayashizaki, Y., Hillier, L. W., Hinrichs, A., Hlavina, W., Holzer, T., Hsu, F., Hua, A., Hubbard, T., Hunt, A., Jackson, I., Jaffe, D. B., Johnson, L. S., Jones, M., Jones, T. A., Joy, A., Kamal, M., Karlsson, E. K., Karolchik, D., Kasprzyk, A., Kawai, J., Keibler, E., Kells, C., Kent, W. J., Kirby, A., Kolbe, D. L., Korf, I., Kucherlapati, R. S., Kulbokas, E. J., Kulp, D., Landers, T., Leger, J. P., Leonard, S., Letunic, I., Levine, R., Li, J., Li, M., Lloyd, C., Lucas, S., Ma, B., Maglott, D. R., Mardis, E. R., Matthews, L., Mauceli, E., Mayer, J. H., McCarthy, M., McCombie, W. R., McLaren, S., McLay, K., McPherson, J. D., Meldrim, J., Meredith, B., Mesirov, J. P., Miller, W., Miner, T. L., Mongin, E., Montgomery, K. T., Morgan, M., Mott, R., Mullikin, J. C., Muzny, D. M., Nash, W. E., Nelson, J. O., Nhan, M. N., Nicol, R., Ning, Z., Nusbaum, C., O'Connor, M. J., Okazaki, Y., Oliver, K., Overton-Larty, E., Pachter, L., Parra, G., Pepin, K. H., Peterson, J., Pevzner, P., Plumb, R., Pohl, C. S., Poliakov, A., Ponce, T. C., Ponting, C. P., Potter, S., Quail, M., Reymond, A., Roe, B. A., Roskin, K. M., Rubin, E. M., Rust, A. G., Santos, R., Sapojnikov, V., Schultz, B., Schultz, J., Schwartz, M. S., Schwartz, S., Scott, C., Seaman, S., Searle, S., Sharpe, T., Sheridan, A., Shownkeen, R., Sims, S., Singer, J. B., Slater, G., Smit, A., Smith, D. R., Spencer, B., Stabenau, A., Stange-Thomann, N., Sugnet, C., Suyama, M., Tesler, G., Thompson, J., Torrents, D., Trevaskis, E., Tromp, J., Ucla, C., Ureta-Vidal, A., Vinson, J. P., Von Niederhausern, A. C., Wade, C. M., Wall, M., Weber, R. J., Weiss, R. B., Wendl, M. C., West, A. P., Wetterstrand, K., Wheeler, R., Whelan, S., Wierzbowski, J., Willey, D., Williams, S., Wilson, R. K., Winter, E., Worley, K. C., Wyman, D., Yang, S., Yang, S. P., Zdobnov, E. M., Zody, M. C., & Lander, E. S. (2002). Initial sequencing and comparative analysis of the mouse genome. *Nature, 420,* 520–562.

Index

© The Author(s), under exclusive license to Springer Nature Switzerland AG 2020
G. K. Pandey, S. Mahiwal, *Role of Potassium in Plants*, SpringerBriefs in Plant
Science, https://doi.org/10.1007/978-3-030-45953-6

Printed in the United States
By Bookmasters